PROSTATE CANCER

SEXUALITY SURVEY RESULTS

SURVIVOR & PARTNER

Dr. Jo-an Baldwin Peters

Dr. Joel Funk,

Court Brooker

So you have had treatment or are considering treatment for prostate cancer. Do you, or will you, have erectile dysfunction as a result of treatment? Are there ways to minimize a potential problem?

<u>Editing…..</u> <u>Colin Peters, Mairi MacKinnon</u>

Cover design and photography……Court Brooker

Graphics/Survey Mechanics…..Court Brooker

Cover Artwork…..Dr. Jo-an Baldwin Peters

All Rights Reserved. No part of this book may be reproduced or transmitted in any form or by any means without written permission from Dr. Jo-an Baldwin Peters. Email: dr.jo.an.p@gmail.com.

Table of Contents

Table of Contents 2

About the Team 3

Preface 4

Prostate Cancer SURVIVOR Sexuality Survey 14

    Introduction to Actual Survey 14

    Purpose of the Study 14

    Study Design 14

    Who Can Participate in the Study 15

    Who will Conduct the Research 15

    What Will You Be Asked To Do 16

    Possible Risks and Discomforts 16

    Compensation / Reimbursement 16

    Confidentiality and Anonymity 17

    Questions or Concerns 17

Prostate Cancer Survivor PARTNER Sexuality Survey 126

    Introduction to Actual Survey 126

    Purpose of the Study 126

    Study Design 126

    Who Can Participate in the Study 126

    Who Will Conduct the Research 127

    What Will You Be Asked To Do 127

Possible Risks and Discomforts	*128*
Compensation / Reimbursement	*128*
Confidentiality and Anonymity	*128*
Questions or Concerns	*128*
In closing	**177**
Resources & References	**179**
Bibliography	*179*
Internet sites cited by PCa survivors: survey respondents	*180*

About the Team

Dr. Jo-an Baldwin Peters is an independent researcher and the wife of a prostate cancer survivor. Her qualifications encompass a BSc in Physiotherapy (Physical Therapy), MSc in Epidemiology and Biostatistics, and PhD in Health Care Administration. Her dissertation focused on how prostate cancer treatments impacted the sexuality of both partners. In addition to her research, she gives presentations to prostate cancer support groups in Canada and the US, has participated in radio talk shows and has published articles for prostate cancer survivors and their partners.

Dr. Baldwin Peters is responsible for getting the online sexuality survey up and running. The survey dealt with sexuality issues that arose after treatment for prostate cancer, and included both survivors and their partners. Her contribution to the project included research, writing, development, cover art, creativity, promotion, spunk and friendship.

Dr. Joel Funk is a certified Urologist, initially associated with the Yavapai Regional Medical Center in Prescott, Arizona. He is now Assistant Professor of Surgery at the University of Arizona in Tucson, Arizona. He got his MD at Northwestern University Medical School and completed his Urology residency in Tucson at the University of Arizona. He is a major in the U.S. Army Reserves.

Dr. Joel Funk was ever present as a source of ongoing and powerful support.

Court Brooker underwent a radical prostatectomy in January 2008, at age 59. Several months later, he attended a local prostate cancer support group and subsequently became the facilitator. The successful collaboration between Court and Dr. Baldwin Peters began after her presentation to his support group.

Court contributed secretarial work, graphics, cover graphic, SurveyMonkey design and data analysis, marketing assistance, research, promotion and his warped sense of humor.

Preface

The authors of the book are the wife of a prostate cancer survivor, a man diagnosed with prostate cancer and a practicing urologist, who bring their own first-hand experience of the effects of this disease. This survey was conducted to provide insight on how prostate cancer survivors and their partners deal with the sexual after effects of the various forms of treatment. If ignored, sexual dysfunction can have a devastating effect on people's lives, leading to sadness, frustration and even despair.

This book presents the results of the survey exactly as provided by SurveyMonkey. The participants' comments stand as we found them, verbatim and unedited except for the exclusion of any items that impinge on the anonymity of the participants. Observations on the results are made by Dr. Jo-an Baldwin Peters and shown in *large bold italics*.

Here's a summary of some of our initial findings (some quite unexpected) dealing with sexuality issues:

- An overwhelming number of responses reinforce the important role of orgasms (achievable without an erection or ejaculation). Exploring the possibility of orgasm without penetrative sex could be a vital component in compensating for and removing stress due to the loss of ability to have normal intercourse.

- The most frequently used methods of achieving an orgasm cited in the survey were mutual masturbation, followed by self masturbation, then oral sex and oral medication. (Of interest, the methods we probably used as adolescents may be the most helpful.)
- There are many ways of achieving an orgasm and this is a matter of personal choice. Individuals are most successful if they communicate and work within their personal comfort zones.
- Resources respondents cited for finding information included, in this order: Internet; physicians; cancer support groups; books; other survivors; library, newspapers, magazines.
- People need information *before* deciding on treatment. In addition, many would appreciate help assessing the reliability, validity and currency of information they find through the Internet and other sources.
- Support groups need to increase their efforts to assist prostate cancer survivors and their partners.

Although the survey did not meet rigid research methodology, Dr. Funk felt that the results could be beneficial in helping numerous prostate cancer survivors and their partners. This survey, as a pre-trial study, could also provide baseline information for future researchers. Much of the data requires further study and analysis. Unfortunately, our group has run out of time, biostatistics expertise and funding. We would be happy to share this data with other researchers or receive funding for a biostatistician or find a volunteer to assist us with data analysis.

The honest and open responses of those who responded to our survey offer valuable information encompassing treatment choices, (and) issues of intimacy and sexuality. From their extensive comments, it was clear that both survivors and their partners were anxious to get their messages heard. We thank all the participants for their honest and insightful contribution, and are committed to continuing to try to get this information out to prostate cancer survivors and their partners.

Some of our ongoing initiatives aimed at fulfilling this obligation include:

- Doing presentations in Canada and the US

- Publication of the complete survey on a shared website — pcainaz.org. The site also features a video presentation by Dr. Jo-an Baldwin Peters and a variety of articles on prostate cancer and sexuality; some of these have also been published on the Us Too website (www.ustoo.org) and in the Canadian *Our Voice* magazine (www.ourvoiceinprostatehealth.com).
- Presentation of our findings to support groups, websites and blog sites on where and how survivors found helpful information as well as their recommendations
- Completion of the data analysis from this survey
- Publication of articles in traditional magazines

Preview of Survey Data

The following charts are a preview of the survey data and briefly summarize some of our findings. The actual summary data follow.

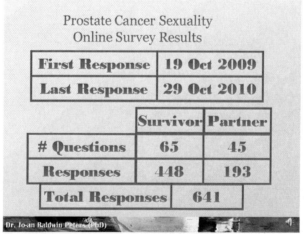

Fewer partners than survivors participated in the survey; this was a pity as it was one of the few opportunities for them to communicate and to validate their viewpoints.

We were able to match 69 couples by cross-matching date-of-birth entries.

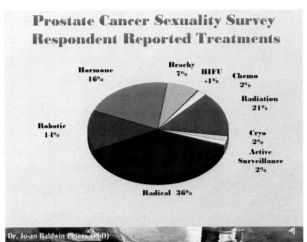

Of the men who completed this survey, 50% underwent prostatectomies (including robotic procedures).

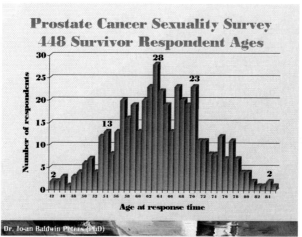

Participants' ages ranged widely, from 42 to 84 years. Examining the distribution, 28 of the respondents were age 63 years; 2 were 42 years; 13 were 54 years; and at the other end of the spectrum, 3 were age 82 years or older.

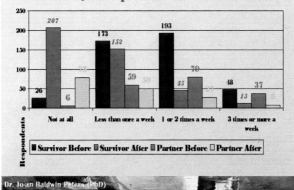

23/9. How often did you have penetrative sex before/after prostate cancer treatment?

After treatment, the frequency of penetrative sex took a "nose dive" compared to pre-treatment figures. When examining the data using respondent percentages there was concurrence between partners and survivors regarding frequency of penetrative sex. However, the partners reported more sexual activity 3 or more times per week than the survivors, but these were not matched partners. This data has yet to be examined using treatment type as the baseline. Question 25 does provide some insight into sexual behavior by age and treatment.

S25. How would you rate your ability to have a spontaneous erection after treatment for prostate cancer? Treatments/Time/Age

Respondents answered "Sometimes" or "Always"

Months	3	6	12	18	24	+24	Resps.
Radical prostatectomy	14%	22%	26%	27%	28%	29%	144
Radical retropupic prostatectomy	13%	20%	13%	13%	20%	20%	15
Radical nerve sparing prostatectomy	25%	33%	40%	46%	42%	38%	48
Robotic prostatectomy	16%	19%	22%	22%	27%	21%	73
Radicals All	16%	23%	27%	28%	30%	28%	280
Brachytherapy/seed implants	66%	68%	73%	66%	56%	56%	41
Cryosurgery/Crotherapy	44%	44%	33%	33%	33%	44%	9
External beam radiation	50%	60%	50%	40%	50%	30%	10
Proton beam radiation	100%	60%	60%	60%	60%	60%	5
Radiation Totals	67%	60%	53%	47%	53%	40%	15
Age 50 and less	31%	13%	19%	13%	19%	13%	16
Age 50-60	23%	30%	33%	31%	31%	30%	122
Age 60-70	43%	49%	54%	55%	56%	54%	144
Age 70+	21%	21%	24%	24%	21%	22%	102

Dr. Jo-an Baldwin Peters (PhD)

Examining the results of the various forms of prostatectomy showed that initially, erectile function was lower compared with other types of treatment. The improvement over a 24- month period, however, was significant, with radical nerve-sparing prostatectomy in the lead.

The radiation group, on the other hand, showed a significant decrease in erectile function over a 24-month period. The number of men receiving radiation was much smaller and these findings need to be followed up with a larger group. This decline may have been due to aging. It must be emphasized that no matter what kind of treatment, it results in significant erectile dysfunction.

Given the fact that there is recovery over time, it is extremely important to exercise the penile tissue. Oral medication should be started early, even if it doesn't result in an erection, and followed up with penile injections (shots) at 3 weeks post treatment. Some surgeons advocate the use of a vacuum pump.

It is interesting to note that the age group that experienced better recovery of erectile function was between 60 and 70 years.

How can this be explained? A more detailed evaluation of these results would be important.

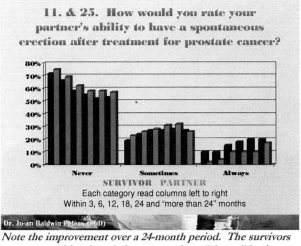

Note the improvement over a 24-month period. The survivors are more positive than their partners about "always" having a spontaneous erection after 24 months. Remember, however, that this is not the matched partner sample.

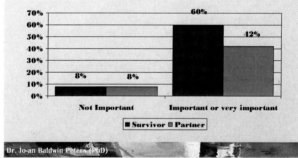

How important is it to experience an orgasm on a regular basis even if it does not result from penetrative sex?

This question is infrequently asked, but it is important for both partners as orgasm is achievable without an erection or ejaculation. In view of this, the question could be rephrased as, "How important is penetrative sex? Will having an orgasm suffice?" Exploring this could be a vital factor in compensating for and removing stress from the loss of ability to have penetrative sex.

33. Which of these forms of assistance resulted in an orgasm with or without penetration?

	Did not achieve orgasm	Achieved orgasm infrequently	Achieved orgasm most of the time	Achieved orgasm always	Response Count	Success Rate
Self masturbation	12	26	85	65	190	94%
Mutual masturbation	7	16	48	25	103	93%
Penile shots	8	8	23	18	59	88%
Oral sex	16	16	31	29	94	83%
Sex toys such as vibrators	4	3	9	5	24	81%
Penile implant	3	2	5	5	18	80%
Pump	19	16	19	9	68	70%
Oral medication (provide medication name in comment box below, if possible)	31	18	29	24	113	70%
Other	1	1	1	0	5	67%
Nothing	11	7	6	6	54	63%
Urethral suppositories	8	3	2	2	18	47%

Dr. Jo-an Baldwin Peters (PhD)

This table identifies methods of achieving an orgasm. They are listed in order of success: self masturbation, followed by mutual masturbation, penile shots and oral sex.

Analysis by age and treatment would add important contextual information.

Prostate Cancer SURVIVOR Sexuality Survey

Introduction to Actual Survey

This survey is for prostate cancer survivors. There is a separate survey for their partners.

The consent form is an agreement to do the survey which deals with the after effects of prostate cancer treatment on the sexuality of both partners.

We invite you to take part in a research study being conducted by Dr. Joel Funk, Dr. Jo-an Baldwin Peters (PhD) and Court Brooker. Your participation in this study is voluntary and you may withdraw at any time before the survey is completed. The study is described below and identifies any risks, inconveniences or discomfort that you might experience.

Participating in the study may not benefit you directly, but we expect to learn things that will benefit others. You should discuss any questions you have about this study with Dr. Jo-an Baldwin Peters who can be contacted at dr.jo.an.p@gmail.com.

Purpose of the Study

Through this study, we anticipate gaining more knowledge about how the after effects of prostate cancer treatment impinge on your sexuality and that of a partner. Recommendations can then be made to treating physicians and support groups, and appropriate materials can be developed in relevant formats and distributed in accessible outlets to prostate cancer patients.

Study Design

This study has been reviewed for content, and ethical and confidentiality issues by a panel of experts. The survey invitation will be posted on several related websites for three to six months. All submissions will be completely anonymous. At no point will your name or other identifying information be requested or acquired. It is our intent to halt the survey when 3,000 responses are received.

Who Can Participate in the Study

You and your partner can participate in this study survey by visiting several websites or receiving an email link. The survey can be accessed/shared by copying the following links and pasting the appropriate one into your browser address field or emailing the link to other known prostate cancer survivors:

Survivor Survey Link: No longer operational as survey ended November 2011

Partner Survey Link"No longer operational Prostate cancer survivors without a partner may also participate and are encouraged to do so. As this survey focuses on men who have received treatment for prostate cancer and their partners, both partners are ideal candidates to participate in this survey. Prostate cancer survivors without a partner may also participate and are encouraged to do so.

Who will Conduct the Research

Dr. Joel Funk is the Principal Investigator of this study together with Dr. Jo-an Baldwin Peters (PhD) and Court Brooker.

Dr. Joel Funk is a certified Urologist initially associated with the Yavapai Regional Medical Center in Prescott, Arizona, and currentlyAssistant Professor of Surgery at the University of Arizona in Tucson.

Dr. Baldwin Peters is an independent researcher and the wife of a prostate cancer survivor.

Court Brooker is a prostate cancer survivor experienced in communications and media.

The study design was a collaborative effort between Dr. Joel Funk,

Dr. Jo-an Baldwin Peters and Court Brooker.

The survey will be administered via the internet.

What Will You Be Asked To Do

The following survey consists of a maximum of 65 questions. Based on a number of trials it is estimated that this will take you 15 to 25 minutes to complete. This is the only task that you will be asked to do. Please note that detailed personal questions will be asked regarding your past and current sexual ability as well as your sexual orientation and MUST BE COMPLETED INDEPENDENTLY BY EACH PARTNER.

Possible Risks and Discomforts

You may experience some emotional discomfort while answering the survey as many of the questions are specifically about your past and current sexual ability. You may skip questions you find too difficult to answer. You may also withdraw from the study at any time.

We will make all publications based on this data available through our website by posting information about how to obtain them as soon as they are published. We will provide copies in PDF format to all who request them.

There are no direct benefits to you for participating in this study. However, our study may help us identify areas where better information is needed and in what format and for whom.

Compensation / Reimbursement

You will not incur any expenses beyond the time it takes you to complete the questionnaire. Therefore, no compensation is being offered.

Confidentiality and Anonymity

The data collected by this survey will be of a personal and sensitive nature, but will not be linked to any identifying information.

All data will be retained on personal computers for approximately two years under password-protected files, at which point the files will be erased.

Questions or Concerns

Any questions or concerns about the study can be addressed to Dr. Jo-an Baldwin Peters at Dr.Jo.An.P@gmail.com. It should be noted that in contacting her by email you will be providing her with an email address that may compromise your anonymity. However, your email address is not linked to the questionnaire in any way.

Therefore, Dr. Baldwin Peters will not be able to connect you with your survey responses unless you provide her with sufficient information to do so.

If you are blocked from submitting or advancing whilst completing this survey it may be necessary to revise or re-enter some data. The system will return you to the question or questions requiring revision. Instructions will appear in red and be marked with an asterisk.

Unfortunately this system disallows fractions and requires dates to be fully completed. Correcting the data will then allow you to submit your survey.

Survey Questions and Results

1. Consent form for the study to identify how the after effects of prostate cancer treatments impinge on the sexuality of individuals and their partners. I have read the explanation about this study. I have been given the opportunity to discuss it with Dr. Joan Baldwin Peters and any questions have been answered to my satisfaction. I hereby consent to take part in this study. However, I realize that my participation is voluntary and that I am free to skip questions and/or withdraw from the study at any time. I also authorize the use of quotations from any of my answers to demonstrate my individual sentiments so long as I am not identifiable. By clicking on the 'agreement' button I electronically sign that I accept the terms presented above. (answer required to continue survey)

Answer Options	Response Percent	Response Count
I agree	99.8%	447
I disagree	0.2%	1
	answered question	448
	skipped question	37

2. Birth dates (at least one birth date required to continue survey) Example: For 14th December, 1930 enter DD=14 MM=12 YYYY=1930

Answer Options	Response Percent	Response Count
Your birthdate	99.6%	446
Partner's birthdate	79.2%	355
	answered question	448
	skipped question	37

3. I currently reside in

Answer Options	Response Percent	Response Count
Current City/Town	97.7%	432
Current Province/State/Other	96.8%	428
Current Country	95.0%	420
	answered question	442
	skipped question	43

4. Relationship?

Answer Options	Response Percent	Response Count
Have no partner	9.7%	43
Have a partner	90.3%	401
If you have a partner, how many years have you been partners?		405
answered question		444
skipped question		41

While 401 men had partners, only 193 partners answered the survey and thereby lost an opportunity to identify their specific problems. Partnership duration ranged from 4 months to 63 years.

5. Your weight (answer one)? (Whole numbers only, no decimals)

Answer Options	Response Average	Response Total	Response Count
Pounds	240.88	102,133	424
Kilograms	81.61	1,877	23
answered question			445
skipped question			40

Average weight 193 lb/88 kg. Weight range 120 lb to 340 lb.

6. Your height (answer one)? (Whole numbers only, no decimals)

Answer Options	Response Average	Response Total	Response Count
Inches	70.11	29,445	420
Cm	159.29	3,823	24
answered question			441
skipped question			44

7. Your education?

Answer Options	Response Percent	Response Count
High School	20.5%	91
College/University graduate	41.5%	184
Post gradauate	34.3%	152
Other	3.6%	16
answered question		443
skipped question		42

Respondents were highly educated; this bias must be considered.

8. Your race?

Answer Options	Response Percent	Response Count
Caucasian	93.3%	416
Asiatic	0.4%	2
African American	4.3%	19
First Nations	0.4%	2
Other	1.6%	7
answered question		446
skipped question		39

9. The following would best describe me

Answer Options	Response Percent	Response Count
Heterosexual	94.8%	419
Homosexual	2.9%	13
Bisexual	1.8%	8
Other	0.5%	2
answered question		442
skipped question		43

This survey consisted mainly of heterosexual men; the questions were unintentionally biased towards this group.

10. What relatives on your father's side have had prostate cancer?

Answer Options	Response Percent	Response Count
Brother	8.4%	37
Father	24.4%	107
Uncle	8.7%	38
Cousin	4.8%	21
Other	2.1%	9
Not sure	24.2%	106
None	39.7%	174
If Other, who?		19
answered question		438
skipped question		47

Q10. Respondent Comments to: "If Other, who?"

1	unsure none I know of
2	Secon cousin once removed
3	Paternal grandfather
4	only child
5	Not sure of some of fathers's brothers
6	No Knowledge of Birth Parents
7	My Grandfather
8	Grandfather (maternal)
9	Grandfather
10	Grandfather

11	Grandfather
12	Grandfather
13	Grandfather
14	Grandfather
15	Grandfather
16	GRAND FATHER
17	Grand Father
18	Father's Father
19	2 br

11. What relatives on your mother's side have had prostate cancer?

Answer Options	Response Percent	Response Count
Brother	1.4%	6
Father	1.8%	8
Uncle	9.5%	42
Cousin	2.3%	10
Other	1.4%	6
Not sure	40.2%	178
None	45.8%	203
If Other, who?		9
	answered question	443
	skipped question	42

Q11. Respondent Comments to "If Other, who?"

1	No Knowledge of Birth Parents
2	Mothers Grandfather
3	Great Grandfather
4	Grandfather
5	grandfather
6	Grandfather
7	Grandfather
8	Grandfather

12. What date was your prostate cancer first suspected, diagnosed and when was my first treatment? (estimate if uncertain)

Answer Options	Response Percent	Response Count
My prostate cancer was first suspected on	94.3%	399
My prostate cancer was first diagnosed on	94.3%	399
My first prostate cancer treatment was on	92.7%	392
	answered question	423
	skipped question	62

The average time between diagnosis and treatment was 3.3 months; range 1 to 16 months. We haven't evaluated this data by treatment, country, state or province and no doubt these factors would show variations.

13. How was your prostate cancer first suspected? (one or more choices may be selected)

Answer Options	Response Percent	Response Count
Rising PSA	79.7%	353
DRE(digital rectal exam)	33.2%	147
Biopsy	30.5%	135
Ultrasound	4.3%	19
Pelvic CAT scan	2.3%	10
Bone scan	1.8%	8
Other (please explain)	9.5%	42
answered question		443
skipped question		42

Rising PSA was the first indication of prostate cancer. This data needs to be evaluated by country to identify differences in diagnostic methods.

Q13. Respondent Comments to "Other (please explain)"

1	With PSA of 2.5, I volunteered for a biopsy study at [...]. One of 11 cores was positive.
2	Weak uring stream
3	Very high PSA First test
4	Urine retention
5	Urinary problems.
6	Urinary difficulty
7	TURP; 2 of 80 samples came back as positive for cancer, gleason score 3+3 = 6
8	Turned out to be BPH
9	Tired feeling brought up at yearly physical
10	The 7th biopsy detected cancer.
11	Spike in PSA to 8.4 as a result of a case of prostatitis.
12	Resection of enlarged prostate -- resulting tissue removed showed some cancer present.
13	rectal exam , suspected node , sent to specialist who preformed biopsy PSA Was in the Normal range for my age
14	PSA was slightly higher than expected. Never really rose aggressively (no "doubling")
15	PSA level of 4.5 at age 40, no other signs
16	PSA = 5.8 - no previous PSA tests
17	Prostatitis with reocurring infections after cipro regimen.
18	Primary Care MD, piggy-backed PSA along with Cholesterol test. Urologist confirmed via PSA, Free-PSA and then Biopsy.
19	Persistent Bladder infections sent me for TURP and biopsy
20	overactive bladder

21	My PSA was rising for over five years. Doctor treated four times with cypro, but never pushed me to needle biopsy. By the time I had the biopsy, the cancer was Gleason 8 and T3. Surgery was not an option.
22	MRI
23	LUTS
24	I was weak and could barely walk. The clinic would not let me come in until my appointment day. I had to get some neighbors to help me go in. When they saw how bad I was they sent me to the emergency room. My spine had partially collapsed. They did tests that revealed my psa was over 500.
25	I was in a 7 year medication study.
26	I had a psa of .7 from age 40-47. It then moved to 102 for ages 47-50. I had always had a ridge on the center of my prostate detected during DRE's. That combined with an increase 3 years before as well as family history, it was recommened that I have a biopsy. 7 samples, 3 from each hemisphere were positive, the one from the ridge was neg. Gleason was 3+3 and Stage was T2B (couln't enter exact below)
27	Haemospermia
28	Had a physical and PSA was checked - very high.
29	frequent urinating, caused doctor visit
30	found on colonoscopy by gastro enterologist..normal PSA ..no DRE for past 7 years by family practice physician
31	Family Doctor Detected Growth on Prostate During Annual Physical Checkup
32	Excessive urine
33	Elevated PSA on first ever test. Yearly DRE's were all normal.
34	During normal blood test for cholesterol
35	During my annual check up a PSA was taken, it was 50.
36	Difficulty passing urine in the mornings
37	Difficult urination became impossible urination, urthera was totally blocked, had to have catheter installed
38	Blood in semen
39	Bleeding in ejaculation
40	beginnning ED
41	2 brothers had prostate cancer
42	1st ever PSA at age 49; no symptoms; nothing detectable on DRE; PSA was 13.2

14. Do you know your PSA, Gleason Grading or Cancer Stage at time of diagnosis?

Answer Options	Yes	No	Response Count
I know my PSA at the time of my diagnosis	403	32	435
I know my PSA at the time of my treatment	328	63	391
I know my PSA now	399	21	420
I know my Gleason Grade	343	64	407
I know my Cancer Stage	261	121	382
		answered question	444
		skipped question	41

Does this data vary by country? It would appear that most men know their PSA and Gleason Grade but not their cancer stage. It is difficult to ascertain whether these grades are similar from country to country.

15. If you know them, please enter your PSA, Gleason Grade and Cancer Stage please enter them here. (Whole numbers only, no decimals, raise to the next whole number for fractions in all cases) (leave blank if you do not know)

Answer Options	Response Average	Response Total	Response Count
PSA at time of diagnosis	16.16	6,527	404
PSA at time of treatment	17.34	6,070	350
PSA now (if less than one go to next question)	7.33	968	132
Gleason Grade	6.76	2,212	327
Cancer Stage	2.03	358	176
		answered question	418
		skipped question	67

16. If your PSA is less than one choose one.

Answer Options	Response Percent	Response Count
.9 to .5	9.1%	37
.4 to .1	14.8%	60
under .1	64.2%	260
Does not apply	11.9%	48
	answered question	405
	skipped question	80

17. What type of treatment/treatments did you undergo? (can check more than one)

Answer Options	Response Percent	Response Count
Cryosurgery/Crotherapy	2.3%	10
Radical prostatectomy	37.5%	166
Radical retropupic prostatectomy	8.8%	39
Radical nerve sparing prostatectomy	20.5%	91
Robotic prostatectomy	20.5%	91
External beam radiation	19.2%	85
Proton beam radiation	1.4%	6
Intensity Modulated Radiation(IMRT)	5.2%	23
3 dimensional conformal radiation therapy (3D-CRT)	4.3%	19
Brachytherapy/seed implants	9.3%	41
High dose rate Brachytherapy	1.6%	7
High intensity focused ultrasound (HIFU)	0.5%	2
Hormonal drug therapy	23.0%	102
Chemotherapy	2.9%	13
Active Surveillance	2.3%	10
None	0.0%	0
Other	3.4%	15
Comment		53
answered question		443
skipped question		42

Q17. Respondent Comments

1	*Zoladex @ 3 month intervals, bicalutamide daily (trying every second day for the last 6 weeks to alleviate breast pain)*
2	*When PSA rose in 2007 it was decided to follow up with radiation and 1 yr Zolodex in 2007-08*
3	*used targeted nutrition, supplements, exercise, reducing toxin exposure, laughter, relaxation, stress control, forgiveness and prayer. Have quarterly psa checks and dre*
4	*Traditional cure*
5	*The surgery went VERY well, with little after affects*
6	*The radiation oncologist and the urologist botched the first treatment and I had to have it done again*
7	*The IMRT was done for two months as the cancer reoccured after six years.*
8	*Surgery followed up by Hormones and radiation*
9	*Still in treatment with 33 0f 45 treatments completed as of Oct 20, 2009.*
10	*robotic laproscopic prostatectomy*
11	*Radical Prostatectomy failed, was followed six months later by external beam radiation*
12	*qtrly PSA for last 3 years..complete work up at XXX June 2009... no cancer found... submit PSA`s quarterly to oncologist*
13	*Presently on 150 mgms Casodex -20 mgms tamoxifen daily*
14	*Phase III Dendreon Clinical Trial*
15	*Perenial prostatectomy*
16	*Pcs, Prostasol, Zometa, Samarium, Vaccine dendtritic cells*

17	On hdt for 7 years
18	Not sure if I am labelled as T3 or T4. Cancer was found in my bones.
19	NON TRADITIONAL WITH VITAMINS, MINERALS
20	Nerve sparing
21	My urologist said he performed a bilateral nerve sparing prostatectomy. I am not sure of the stage, but I think it was T2. The pathologist report revealed 3 focci, one of them very close to the edge of the gland. The DRE could not detect any abnormality because the tumors were on the other side of the prostate where it could not be palpated.
22	minimal radiation was done using EBRT and described as "palliative"
23	Lupron about eight years later. Currently on Lupron 3. I highly recommend it for a number of reasons.
24	Lupron
25	LRP none nerve sparing
26	Laprosocpic
27	Laparoscopic nerve sparing radical prostatectomy
28	laparascopic surgerical procedure (LRP) - nerve sparing both sides
29	In fact wasn't "nerve sparing" but was supposed to be.
30	In 12/1001 I was injected with a 3 month Lupron dose in to my left buttock.
31	I received salvage IGRT in 2008
32	I am on my 6th, one year on cycle of triple hormonal blocking, with each off cycle lasting about 18 months.
33	Hormone therapy was for 9 months prior to surgery in 2001. Currently on Zoladex and Casodex
34	Hormone followed by Radiation
35	Hormonal Treatment commence in May 2009 when PSA had risen to 22 (It was 4 after surgery in 2005
36	Hormonal therapy was used to shring the size of the prostate before IMRT.
37	Hormonal first and for 3 mo's following surgery.
38	Hormonal drug therapy was before surgery
39	HDR, removable seeds
40	Focal cryotherapy (right side only)
41	First dose Lupron..08/2009
42	Computerb assisted,
43	Clinical Trial NDGA
44	Chemical farilure was immediate post RPP, HRT failure May, 2009, currently on HD Ketoconazole/Hydrocortisone
45	Casodex, plus had been taking Prosar since 1991
46	Brachyterapy and external beam with apparent cure for 10 years. Now recurrent so tried cryo but failed so am on Hormone blockade with pulmonary mets
47	At California Endocurietherapy Cancer Center, monotherapy.

48	Also, the resection for enlarged prostate -- from which the detection of cancer came.
49	After radical RRP in April 2001, had positive margins, and residual and rsing PSA; had 3 months of daily 3-D Conformal radiation (67 gray?) Oct through Dec 2001; PSA undedectable since then!! Latest tisi month was >0.01!!
50	after radical prostectomy the biopsy showed it had spread to the seminal vesicle and was at outer margins so started lupron shots and 7 weeks radiation
51	After 5 years on Lupron, had bilateral orchiectomy. Cheaper over long run and more direct way to reduce the testosterone. Not offered as option, I had to ask for it.
52	120 seeds iodine 125
53	1. Horm. drug. therapy, later Chemo and Radiaton for one Bonemetastase; Other: Ketoconazole Triamcinolon

We did not look at this data by country where the respondents lived. Treatments in different countries can and do vary, but we would require further support to analyze the data.

18. Smoking cigarettes and alcohol consumption?

Answer Options	No	1 per day	2 per day	3 per day	4 or more/day	Response Count
Have you ever smoked cigarettes?	219	29	13	13	161	435
Have you ever been an alcohol drinker?	107	177	74	32	27	417
Do you smoke cigarettes now?	407	3	2	2	21	435
Do you drink alcohol now?	159	167	55	22	11	414
Comment						138
					answered question	442
					skipped question	43

Q.18. Respondent Comments

1	yes I drink alcohol once a week (3 drinks)
2	Was three beers/wines per week, now one per month.
3	Very occasional alcohol use
4	Very light drinker; might have 1 alcoholic beverage per month. Was very heavy smoker; have not smoked for 9 years.
5	Usually Red Wine....but not always daily. Probably 2-3 small glass per week at most.
6	Two or three beers a week. Depends on how much hockey I play.
7	Trick question? Have previously smoked - quit 1992. I drink occasionally but much less than 1 per day. - no options for these answers above
8	This question is confusing. For alcohal does it mean 1 drink per day? If it is less than one there is no answer to give. Not even a Yes.
9	The questions do not mention less than one drink a day. I may have a glass of red wine once a week or so.
10	That is 1 to 2 ounces one time a week.

11	stopped smoking July 5, 1973 1-2 glasses of chillable red, when available
12	Stopped smoking in 1996
13	Stopped smoking around the year 1990 until about the year 2000. Stopped again around 2001
14	stopped alchohol 31 yrs ago, cigarettes 10 years ago
15	Socil drinker
16	Socially ,,,,,,one drink per month
17	social smoker, quit 30 years ago
18	Social drinking only. Once a week.
19	social drinking only
20	Social drinker; maybe 2 - 3 per month
21	Social Drinker, do not drink everyday
22	Smoked one+ pack a day from age 14 to 31.
23	Smoked from 19 years to 31 years Stopped alcohol cosumption at 65 years
24	smoked for 15 yrs I quit smoking in 1980
25	Smoked 2 ppd 1963-1967
26	Red wine
27	Recovery Alcoholic, 29 years sober
28	Really, I am trying to quit.
29	re alcohol, less than 1/day
30	Quite smoking when I was 26. Drink alcohol once or twice a week
31	Quit smoking over 40 years ago and smoked less than one pack a day. Drink moderately in social situations, usually no more than a drink or two a week.
32	Quit smoking more than 50 yrs ago--never addicted.
33	Quit smoking in 1974. Smoke around 1 cigar a day now and since 1993
34	Quit smoking 35 years ago.
35	quit smoking 30 yrs ago
36	Quit smoking 25 years ago
37	Quit smoking 20 years ago, occasional beer.
38	Quit over 30 years ago.
39	Quit light smoking 30 years ago
40	One small glass of red wine in the evening. Never more, and nothing else.
41	One beer or glass of wine a day. Have not smoked in 25 years.
42	Once a month
43	On average, fewer than 1 per day. Perhaps 2 per day Saturday and Sunday
44	OCCASIONAL DRINK ALCOHOL - SOMETIMES A BEER OR TWO SOMETIMES A MIXED DRINK BUT NOT EVERYDAY
45	Occasional beer, or wine with dinner 2-3 times per month.
46	Occasional alcohol only on weekends. Less than 1 per day but you do not list that option.
47	much less than 1 daily

48	MInor experimentation with cigarettes when I was a teen.
49	less then one/day maybe two a week.
50	Less than one drink per day
51	less than 1 per day, about three drinks a week
52	less than 1 drink/day
53	less than 1 drink of alcohol per day maybe 2 per month
54	It's less than one a day. Usually 1-2 times a week.
55	I'm a brewer!
56	I smoked marijuana from 1972-1981 and 1991-2005.
57	I smoked for 3 months 41 years ago. I drink less than 1 drink per day (3 drinks/week)
58	I smoked for 23 years and it is now 36 years since I completly stopped smoking
59	I smoked 34 years ago. Yes I drink maye one or 2 drinks a month
60	I rarely consume alcohol and usually only one drink per ocassion
61	I quit smoking in 1970.
62	I quit smoking in 1970.
63	I quit occasional smoking when I was 14 years old and have consumed alcohol less than 6 drinks monthly.
64	I occasionally drink wine at dinner but certainly not one a day.
65	I occasionally drink alcohol. I quit smoking in 1967.
66	I might have one drink once or twice a month.
67	I may have 4 or 5 drinks a year
68	I have one and sometimes two glasses of wine with my evening meal.
69	I have occassional drinks
70	I have drank alcohol previously and now. Maybe 1 or 2 drinks per month
71	i have an occasional alcoholic beverage. maybe 1 or 2 per week on average.
72	I have a glass of wine or a beer 4 or 5 times a week with dinner. I have not smoked since 1970.
73	I had smoked for 9 years 50 cigarets daily. I haven't had one in more than 48 years
74	I gave up smoking on April 11, 1971
75	I DRINK SOCIALLY, NOT EVERY DAY, OCCASSIONALLY
76	I drink one or two drinks per month.
77	I drink mostly wine with dinner, and an occasional beer or an occasional mixed drink (like scotch and diet-seven).
78	I drink less than 1 drink a day, maybe 2 a week. Will start a stop smoking program in a few weeks.
79	I drink beer and wine occasionally -- about once per week or less.
80	I drink alcohol socially...1 or 2 a week
81	I consume wine socially only on weekends or special occasions
82	I consume approximately 2 to 4 cans or bottles of beer per week. No hard liquor.
83	I consider myself a social drinker having maybe two to four in an evening. this may occur once every three to four months

84	I chew Skoal smokeless tobacco now, and have for the past 15 years
85	I assume in reference to cigarretes the questionnaire means 1 pack per day, 2 pack per day etc. I quit smoking 22 yrs ago and had smoked intermittently for the previous 10 yrs. I would smoke for a year or two, quit for a year or two etc.
86	I am alco;holic . Joined AA 48 years ago and have been clean and sober ever since.
87	HAVEN'T SMOKED IN 30 YEARS
88	Haven't smoked in 17 years Alcohol comsumption once or twice a year.
89	Haven't smoked in 10 years. Smoked a pipe.
90	Have not smoked or drank for 35 years
91	Have always drunk real ale, less as I have got older, very little wine and like smoking never had spirits
92	glass of wine with dinner
93	Glass of wine with dinner
94	Gave up smoking in 1983
95	father and mother smokers..father recovered alcoholic...brother alcoholic..both deceased
96	Drinking way down due to Keto liver side effects
97	Drinking varies...three at time is max, 0 is min.
98	Drink: approx 2 per month, then and now.
99	Drink only on weekends
100	drink on weekend, 4 drinks
101	Drink less than 1 per day - before & now
102	Drink beer and wine, sometimes go a few days without.
103	drink alcohol on ocasions only
104	Don't drink every day
105	Do not drink every day . some days 2 drinks some days none and of course days of zero drinks
106	Do mean packs or cigarettes. I smoked about a pack a day. I don't drink alcohol every day
107	discontinued both over 25 years ago
108	Consumtion of alcohol is like one bottle of wine every two weeks >> Cigarettes quit 28 years ago or more
109	Cigars not Cigarets Alcohol once per week
110	cigaretts is 1 pack per day
111	cigarettes-4yrs. alcohol---10yrs.
112	Cigarettes: 1953 - 1968 Pipe: 1968 - 1983. Quit in the Great American Smokeout of 1983.
113	Average cigarette consumption over a week is 2-3 per day. Alcohol is 1 drink per week on average.
114	Amount of is more important then amount. In my case I have regularly, 12 ounces of wine/day (small glasses).
115	alcohol--probably 2-3 drinks per month before and now
116	Alcohol, less than one per day

117	alcohol, but less than one per day -- mostly a drink or two on the weekend, or a glass of wine when out to dinner
118	Alcohol one or two drinks per week
119	alcohol is only 2 or 3 times per week
120	Alcohol is less than 1 per week, more like 1 per month and sometime not that
121	alcohol is basically wine with dinner.
122	Alcohol consumption occasional; less than one per day
123	alcohol consumption is more like 2 a week
124	Alcohol consumption is infrequent and usually limited to two glasses of wine.
125	alcohol consumption is actually less than 1 per day but not 0
126	Alcohol Consumption 1 or 2 per week
127	alcohol consumation is hit and miss, not a regular thing
128	Alcohol before and since approximately 1-2 per week or less
129	Alcohol about 1 drink per week now Cigs 30-40 per day when quit (1978)
130	About 2 drinks a week and 2-3 beers a week
131	6 oz of red wine 2-3 times a week; nothing else
132	4 cigarettes? It used to be about 1 pack a day..
133	2 glasses of red wine-Medicinal only--LOL
134	2 drinks per month, on average!
135	2 - 3 glasses of wine a day or 1 - 2 Vodka tonics a day
136	1 per month maybe
137	1 or 2 times a week 3 -5 oz total per occassion
138	<1 per day for alcohol, but I don't abstain

It is interesting that a large group of men felt it was important to elaborate on their answers.

19. Please check the following conditions that apply to you?

Answer Options	At diagnosis	After treatment	Response Count
Diabetes	28	30	45
Depression	38	65	78
High blood pressure	151	111	181
High cholesterol	175	120	201
Heart problems	41	46	61
Enlarged prostate gland	135	15	137
Had a vasectomy	176	31	176
No medical concerns	101	48	106
Bladder cancer	2	2	3
Other cancer	25	11	32
Other (please specify)			67
answered question			427
skipped question			58

Of interest is the increase in depression after treatment.

Q19. Respondent Comments to "Other (please specify)"

1	very mild hypertension
2	undifferentiated connective tissue auto immune disease
3	Type 2 diabetes
4	Thyroid
5	testicular 1987, surgery removed 1 testi, no recurance
6	Testicular
7	Stroke X 2, 2004
8	squamus cell, basil cell
9	squamous skin cancer, pre-cancer lesion in bladder
10	small intestine resection 1995
11	skin, blood pressure and cholesterol controled w/meds
12	skin cancer
13	Skin cancer
14	skin (basal)
15	S/P Melanoma 1987 Rt. Lateral Thigh Level 1 Grade 2
16	Rheumatoid Arthritis diagnosed 1 yr. prior to treatment
17	Restless Leg Syndrom
18	Psycho/marital Therapy - I am Gay - my wife is straight
19	Prostate Cancer
20	Previou surgery to remove thyroid(Cancer)
21	pre-diabetes now
22	Peripheral Neurapathy, essential tremor, gout, hypo-thyroid, Hiatal Hernia,
23	Pacemaker
24	OAB
25	No after effect
26	MVP after treatment
27	multiple sclerosis
28	moderate cholesterol 160 - 220 varies
29	Lymphoma cancer in March 2004
30	Lower back problems (2 surgeries--micrtodiskectomies--at L4/L5)
31	left kidney gone; metastatic to left obdurator nodes; left ureter obstructed
32	I was diagnosed with ED at age 49. Used viagra and cialis from then until now.
33	I have a Urinary Implant
34	hypothyroid
35	Hypertriglyceridemia; skin cancer (face)
36	hot flashes
37	history of melanoma 10 years prior
38	Heart problem refers to a recently diagnosed bicuspid aortic valve. It may or may not ever need treatment. In addition cataracts and osteoarthritis in the lower back.
39	head and sholderaccuretic.
40	Gout
41	Glaucoma in one eye

42	glaucoma before and after ; acid reflux before and after
43	gerd...high cholesterol
44	Gerd
45	Factor V Leiden blood condition
46	Esophageal Achalasia
47	Diabetes type 2 - use metformin
48	Crohns
49	Colon Cancer in 2008. Surgically removed.
50	CLL
51	CHRONIC SINUSITUS, SLEEP PROBLEMS
52	Cholesterol 2008; Arthritis...
53	Border line diabetes, now under control
54	Blood pressure and cholesterol were and are being treated and were under control at time of diagnosis
55	Blood clots (from Factor 5 Leiden
56	BASIL CELL, SQUEAMUS CELL
57	Basil cell skin cancer on my face
58	basic cell carcenoma of nose
59	basal carcinoma of the skin years ago successfully treated
60	atrial tachacardy
61	at fibration
62	asthma long before diagnosis
63	Arthritis
64	Also have toxicencplopthy due to exposure to work place chemicals off on disability 10 years
65	after treatment I have pyronies disease
66	Acoustic neuroma excised 12/1/1987
67	"Hairy Cell" Leukemia

20. Do you take medication for any of the above or other conditions?

Answer Options	Response Percent	Response Count
No	31.3%	135
Yes	68.8%	297
Please list those medications		289
	answered question	432
	skipped question	53

A high percentage of survivors are on medication. Could this be a possible factor contributing to erectile dysfunction in this study group?

21. Prior to your diagnosis of prostate cancer did any of the medications you took affect your ability to have an erection sufficient enough for penetrative sex?

Answer Options	Response Percent	Response Count
No	88.3%	384
Yes	4.8%	21
Do not know	6.9%	30
if yes? How? Explain..		36
answered question		435
skipped question		50

Q21. Respondent Comments to "If yes, how? Explain."

1	Without meds - no erection. With meds - the classic erection!
2	Was not taking before diagnosis.
3	Took viagra
4	then partner stopped sex in 1997, for no reason I can think of; for some reason I cannot explain, I did not notice; In workup for angina, blood was taken and PSA was run=4.7; result was not known to me for 10 years.
5	The Lisinopril affect the ability to achieve a full erection. I would take Viagra before sex and would have a strong erection and was able to have penetrative sex.
6	the effects of prozac are well-known
7	Prozac-used Viagra
8	prozac- longer lasting erections
9	Prozac
10	over the course of four years prior to being diagnosed I was treated for an enlarged prostate and was prescribed several different drugs to include (cialis, viagra, terazosin) nothing helped. During this period an ultrasound or biopsy was never administered.
11	occasional, slight reduction due to BP Meds
12	No problems worked as it should
13	No medications involved in sexual activity.
14	N/A.
15	MS caused erectile dysfunction
16	It did change my staying power.
17	I was slowing down. Benicar is not usually supposed to have that side effect, but blood pressue medications, in general, may have.
18	I took Viagera prior to surgery, but it does not work now.
19	I had significant erectile problems before prostate removal. I found that I am severely allergic to Viagra and Ciallis.
20	I did not take any of these medications prior to my treatment
21	I did not take any of these medications before my diagnosis for the CaP
22	I always felt that the blood pressure medication affected me.

23	Have had some ED problems for past 13-years. Probably caused by High Cholesterol according to Urologist. Could have also been physiological as my wife went totally "Non-Sexual". Reason not known.
24	happened simultaneously
25	Had to use viagara
26	Gradual loss of stiffness
27	frequent ED. Treated with testosterone injections and cialis.
28	ED
29	Diminished desire and ability to have erection, not sure it was due to the medication however
30	Did not take any of these till one 16 months after prostate surgery. Had a TIA 16 months after surgery and began meds at that time. I feel that the issues caused by prostate cancer had a direct influence - stress, apprehension, etc.
31	Did not take any medications prior to diagnosis
32	Citalopram affected errections
33	But had ejaculatory delay from anti-depressants. I'm divorced. Had not had intercourse for several years pre-PCa treatment or since. But erections during masturbation were sufficient for penetrative sex before treatment, but generally not since without Viagra.
34	Blood pressure issues made erection unreliable, worked okay some of the time
35	Being a diabetic, I was already having ED issues. I could have an erection sufficient for penetration, but wasn't the best anymore.
36	Based on facts that I now understand, I now believe I have suffered from ED most of my life.

In question 21, 8% of the survivors included comments. On initial review, medication could play a role in the development of erectile dysfunction both prior to and after treatment.

22. Which of the following statements best describes your urinary control?

Answer Options	Total control	Occasional leaking	Frequent leaking	No control	Response Count
Before treatment	396	34	7	3	440
After treatment	120	221	69	28	438
In the past month	184	197	38	9	428
Comment					102
				answered question	442
				skipped question	43

50% of the survivors noted occasional urine leakage; this is a significant number. Of the men who answered this question, 27% had no problems, 15.6% had frequent leakage and 6.3% had no control. This data needs to be further analyzed by treatment type.

Q22. Respondent Comments

1	Would not risk a really full bladder as it might leak but I feel to have full control
2	WHEN TWISTING; HEAVY LIFTING.
3	Weaning off Flomax
4	Weak stream and not leaking have been my issue.
5	wake up several times per night to urinate
6	Very small amounts of leakage, but occurs daily
7	Very rare occasional leaking, a drop or two.
8	Very minor. No need for pads
9	VERY MINOR WHEN STRAINING
10	Very minor leaking, almost never
11	Very minor leaking
12	very brief leak after cryo
13	usually only a small amount after emptying.....when I'm in a hurry
14	Using associated with standing quickly after setting
15	Use one pad per day
16	Urinate frequently. About every 2 Hrs.
17	TURP resulted in some stress incontinence. If I do any walking of any distance, I need to wear a pad. Otherwise, no leaking, even with other physical activities
18	Total control for 9 years after initial incontinence due to RP. Had scar tissue blockage treated with surgery in the doctors office. 5 yrs ago had a procedure to clear out scar tissue at the site of the anastomosis prior to a penile implant and it left me partially incontinent (dribbles)
19	total contol at nine weeks post op
20	there is and has been no control at various times
21	The UI is really irritating, in my opinion. Mainly occurs when lifting and when trying to have intercourse.
22	Surgery caused some incontinence, then radiation therapy has slowly caused ess. total incontinence
23	Stress Incontinence, became worse after radiation
24	Slight weepage, do not require incontinence pads
25	SLHT and low testosterone causing incontinence
26	Seems to vary, even doing the same activities
27	restricted flow
28	Required one pade per day after treatment of 10 weeks only
29	Require an underwear liner
30	Problem with leaking is a dribble after urinating
31	Only when tired
32	Only occurs when I am very tired or after 3+ alcoholic beverages
33	Only minor stress leakage
34	only miner leaking following urination
35	OK 5 weeks after surgury
36	occasional under stress
37	occasional poor control and high frequency

#	
38	Occasional leaking = dribbling into underwear
39	Now only with abdomen pressure
40	now have artificial urinary sphyncter.
41	Nocturnal urination 1 -2 times.
42	No control immediately after catheter removed, for about a week.
43	No change as of yet
44	never a problem
45	Mostly under control at about 7 weeks post RP
46	Mostly stress leakage. Just a small amount
47	more than one drink makes it worse, lifting can cause leaking, tiredness
48	more of a problem with alcohol
49	Male sling installed 2/2009 Failed to work made worse
50	Leaking only when coughing hard and when bladder is very relaxed during foreplay
51	Leaking occcurred for 2-3 months after treatment
52	Leaking may be a bit worse after radiation treatment
53	leaking is insubstantial and does not require pads
54	Leaking improving.
55	leakage only occurs with sex arousal
56	Leakage is minimal and not a distraction.
57	Leak when I contort my body in certain ways
58	It is10 yrs, since radiation and 6 yrs since implant
59	incontinence has declined significantly
60	Implanted AMS 800
61	I'm down to using a folded Kleenex for a pad.
62	I wear 1/2 pad /day
63	I wear "Depends" on a daily basis to prevent leaking.
64	I was very fortunate in this regard
65	I was completely incontinent for 4 months after surgery; at that time I was prescribed imipramine, and continence has been manageable since.
66	I relate this to BP med
67	I regained bladder control about 6 mos. after radiation
68	I now have an artificial sphincter which sort of works
69	I now have an AMS800 artificial urinary sphincter
70	I installed artaficial sphinctor four years ago.
71	I have to urinate at least 4 times in the night
72	I have pretty much total control now, unless I get a bad cough due to my asthma then I have some occasional leaking.
73	I have about 99% control. Only some "dribbling."
74	I had an implant, sphinter valve, installed in 2006
75	I consistently leak at a rate of .2 to .4 ounces per day
76	I am dry during the day but have enuresis about once a week
77	I am close to "social continence," but exertion as minor as walking causes problems
78	Have total control 99.9% of time

79	Have an artificial sphincter
80	Have AMS800 artficial sphincter installed 1yr.
81	Has continually improved no leakage now, just sometime an ergency to go. and more frequent
82	had stress incontinence for one month after cathiter removed
83	Had several injections of cologan
84	Had an artificial sphincter installed 8-09
85	Had a TURP and 2 explority ops. to find source od blood in urine
86	had a male sling installed
87	had a " sling " inserted
88	Had 2 bladder opening surgeries post rad prostatectomy
89	Difficulty with urination now.....3 months later
90	depends on activity eg no leakage when sleeping
91	Complete blockage since March 2009. Self catheterisation used as the exclusive method for urination
92	But can't wait long after I feel the need.
93	Blockage of urinary flow resulted in catheter and resection.
94	Before treatment it was frequent urination not leaking.
95	Bad question, I had urgent urination feelings
96	An occasional drip on completion but no leaking before
97	After treatment, I urniate at time of climax.
98	99.5% of pre-treatment control. Not perfect, but pretty darn good
99	98% continent after surgery and within a few weeks was "dry"
100	100% continent after catheter removal - Dr puts it down to perineal technique. Good success with most patients
101	* weeks post-op began to get control of my incontinence
102	"social incontinent"...well controled with pads & Cunningham clamp, and male sling.

In question 22, 23% of survivors included comments. A fairly large number of respondents have or have had some form of bladder dysfunction.

23. How often did you have penetrative sex before prostate cancer treatment?

Answer Options	Not at all	Less than once a week	1 or 2 times a week	3 times or more a week	Response Count
Before treatment	26	173	193	48	440
After treatment	207	152	45	13	417
				answered question	441
				skipped question	44

This question was answered by 98% of the participants. Before treatment 6% of the men were not having sex at all and 39% less than once per week. After treatment almost 50% (49.6%) were not having sex at all and 36% less than once a week. Pre-treatment 41% of the men were having sex 1 or 2 times a week and post treatment this dropped to 11%. Finally pretreatment sex 2 or more times per week was reported by 11% of the men and this dropped to 3% after treatment. This pre-treatment and post-treatment sexual activity difference identifies the problem that most men experience and find impinging on and diminishing the quality of their life. This is the fear factor that most men face when they have to deal with prostate cancer. This data needs to be analyzed by treatment and age. We are still waiting for help with the data analysis.

24. Were you innovative in your sexual practices (did not only use the missionary position)?

Answer Options	Traditional	Innovative	Response Count
Before treatment	127	301	428
After treatment	112	181	293
		answered question	428
		skipped question	57

It has been postulated that couples who are more innovative in their sexual practices are more able to deal with and solve post-treatment sexuality issues.

25. How would you rate your ability to have a spontaneous erection after treatment for prostate cancer?

Answer Options	Never	Sometimes	Always	Response Count
Within 3 months	295	77	40	412
Within 6 months	250	93	38	381
Within 12 months	205	94	52	351
Within 18 months	175	87	57	319
Within 24 months	159	84	60	303
Longer than 24 months	157	81	58	296
Comment				132
			answered question	439
			skipped question	46

Again, this data has not been analyzed by treatment modality and age.

Q25. Respondent Comments

1	works fine with Viagra
2	with Viagra help
3	With viagra during this time period - now nothing after cryo.
4	With Cialas only. Without never (up to 30 months - present time).
5	Will have a partial erection a few times a week.

6	Wife is still smiling and sexually happy
7	Wife has health problems the preclude intercourse. She contracted Fibromyalgia during my treatment.
8	Very slight enlargement with stimulation
9	Very sensitive issue: I was 49, my wife 36, at diagnosis. Post-op, I immediately asked about Viagra. My Dr's were dismissive, "we don't recommend it for at least 6 months. Be patient, it will take at least a year for the nerve tissue to grow recover. Ultimately, I lost over an inch in length and 20-30% circumference. Now we know that V. should start very soon after surgery; generally, the sooner the better.
10	very sad
11	Very disappointed, MD quoted 50% or better
12	Very depressing. I abosolutely hate my condition.
13	Very depressing!
14	various methods were used to enhance an erection to include vacuum therapy and caverject injection therapy
15	used viagara. cialis. pump and penile injections..remained impotent..due to high grade of tumor no nerve sparing was done..now have a penile implant
16	Unaided spontaneous erections are still almost nonexistent, at the 10 year point. Viagra taken daily (~50mg generic) produces morning erection, but very little otherwise without significant manual stimulation.
17	UI sling installed has had negative effect on erection
18	Trimix Injections work; Cialis did not.
19	Tried three different brands of pills...none worked Tried vacuum therapy....total failure
20	took nearly 5 yrs
21	There has been continual improvement over all 11 years.
22	The involved nerves were competely severed at surgery
23	Still in treatment but sexual desire and performance appears unchanged.
24	Still have to use viagara
25	Still early. Even Viagra is not effective to help me. I'm told I should return to near normal after Lupron is done.
26	Started hormonal therapy in November 2006 from then no erections
27	Spontaneous in sleep only - Requuire textural always now, but can usually get an erection with Cialis - not as strong however
28	spontaneous erections after surgery were and still are not 100%--they are not by themselves enough for penetration and require some foreplay OR meds like Cialis. With Cialis, 100%, just like before treatment.
29	Sometimes, but they do not last more than a couple of minutes and then only during sleep,
30	Sometimes, but not a strong erection and doesn't last long enogh for penetration

31	Since surgery, I have not had an erection. (I am currently starting to try injections ... but so far the injection has only helped maintain a hardon that does not last for more than a few minutes.)
32	Since medicated abiltiy to have a spontaneous erection is never and libido has been lost Answers to Question 26 & 33 are prior to hormone deprivation
33	Sexual experience is better now then before having cancer
34	Several times.......not full erections......perhaps ¼
35	RP left me totally impotent. Used vacuum unsuccessfully, then penile injections (trimix) successfulley for 9 years. Had a penile implant (AMS 800 3-Piece) 5 years ago. Happy with it.
36	Question #23 is answered from the point of view of opportunity not capacity to have sex. Spouse recently died after a long illness!
37	Post treatment erections ALWAYS require stimulation to the penis after a RP
38	Penetration is only accomplished with a shot or a pump.
39	only with injections or pills
40	Only with a ED pump
41	ONLY PARTIAL AS LOST ONE NERVE IN SURGERY
42	only out 6 months now
43	Only once or twice - partial erection for a minute - never had a chance to see if firm enough for penetration.
44	Only occasionally will I have a nocturnal erection and it's usually not firm but it's an improvement over nothing
45	Only been 11 months since surgery
46	Only 4 months post-treatment
47	once on hormones --who cares!!
48	On one occasion, I did have a partial erection - about 2 years after surgery
49	on 2006 had a penile implant
50	NOT YET
51	Not strong enough for penetration. Penis is very thin at the base.
52	not quite yet a month since surgery
53	Not always as hard as it used to be.
54	Nocturnal only
55	Nocturnal erections only
56	No problem
57	no erections, spontaneous or otherwise
58	Nil sex with partner -penetrative or otherwise - during 7 years prior to treatment. Nor since. Biggest problem in life since about 1954 always was avoiding erections and relief thereof. Doubt if bromide would've helped. Nothing would've. Despite urologist advice to contrary, didn't believe prostatectomy would affect. I was wrong!
59	nerves cut during surgery. was told that if that was the case, I couldn't have erections. but I am getting them now. weight loss, exercise, diet, no alcohol helps.
60	Nerve sparing did not work for me.

61	need to use Viagra
62	Need an oral medication such as Viagra in order to get an erection
63	My treatment was 11 months ago, so I didn't respond to the questions referring to more than 12 months. You may want to include a "N/A" (not applicable) column that would apply to similar situations.
64	my treatment was 10 months ago.
65	My problem is lack of desire and wife's lack of desire as well. I have a very small penis. I also have knee problems which makes sex difficult.
66	My Dr. started me on Livetra after three months, I took it for assistance. In the past 14 to 18 Months have not had to use anything.
67	Minor enlargement
68	less than three times in 3 years
69	I've had less than 6 full erections since my operation
70	it's only been 3 weeks since surgery
71	It's been 11 years since my treatment, in the last 3 or 4 years I haven't been able to have spontaneous erections or penetrative sex with my wife.
72	It required about 4 years before I had nocturnal and spontaneous erections
73	it has only been 4 months since the operation
74	insufficient for penetration even with use of Viagra
75	I'm not quite at 6 months yet. I'm a gay man. Some of these questions don't mean quite the same thing for us.
76	I use 20 mg Cialis now, plus a "penis ring" after erection; excellent results; previously used Viagra, then Leveitra, but got very bad headaches; still always able to have penetrative sex. My current and previous 2 relationship partners, and I, have mutually also highly enjoyed oral sex (usually not simultaneously). I always enjoy giving oral sex, and gladly continue for as many orgasisms as my partner wants to have. My experience and feedback has been that clitoral orgasisms are more powerfull for my partner, and internal sex orgasisms are not a sure things, nor as powerfull. My previous partner also enjoyed penetrative sex with a dildo (which she provided and I gladly used) after mutual oral sex. I have been with my current partner for 6 months. I'm always 'up front' about my situation when starting to date a new person.
77	I tryed Viagera, I tryed the pump (which my wife did not like) and I tryed tri-mec (and ended up in the hospital with a long term erection (I think the Dr. prescribed too large a dose) and not I just give up, without an interested partner.
78	I tried vacuum systems with no success. Injections worked.
79	I take nitric oxide almost daily for weight lifting but I also factor in that is helps, and I feel quite confident that it does, help me have more erections and definitely morning erections after 18 months without spontaneuous erections.
80	I have not had an erection since my operation no matter what we try. 9 i will not take Cialis or other drugs because of the fair of the effects.

81	I have had to use 2.5 mg of levitra and that works for me over the last 18 months, but only spontaneous erection was possible without pills after a year.
82	I have erections when I awake but it is very rare. I started getting them 3 yrs after treatment
83	I have been on Lupron most of the time since July 2006, so sex and erections simply aren't possible.
84	I have a friend whom we see each other on occasion but not a full time partner so unsure how to answer some questions in this area. No longer have a friend / partner at this time.
85	I hadn't slept with a woman in 15 years. That is certainly less than once a week.
86	I had surgery 3 weeks ago
87	I had penile implant surgery May, 2001
88	I had a penile implant installed in 2006
89	I believe that the Lupron injection had a devestating effect on my health, breast development, impotence, loss of muscle mass, extreme hot flashes for 6 mos. post injection.
90	I am within 3 months of surgery.
91	I am completly impotent
92	I am bisexual. My wife and I had not had intercourse for several years prior to diagnosis. I masturbated frequently with no problem having an erection.
93	I am 2 months post op. I am able to have a semi hard erection without Viagra and hard with Viagra
94	Hormone therapy influenced this
95	He has not been a diamond cutter in years but viagra, etc. make him semi-hard
96	HAVE TO PLAN AHEAD; BE WELL RESTED; NO STRESS FOR BOTH OF US.
97	have not had an erection after my surgery
98	have no sense of touch in penis after operation
99	have been losing ability to obtain an erection for the last 18 months
100	Four years of LH-RH has shrunk penis in length and girth, shrunk testes to 1/2 size, eliminated all semen, seminal fluid, etc. None of these was ever mentioned by physician. Scarring of cava cavanosa was marked and erections are of short duration, especially with attempt at penetration, where vaginal pressure reduces erection.
101	Erections are seldon capable of penetration, and never spontaneous.
102	Erections are not as firm as I would like them. So, I take a pill.
103	erections are not as firm as before surgery
104	erection only with penile injection of trimix
105	ED medications helped
106	Divorced 2 years before diagnosis, answers are for my 1st wife. Had spontaneous erection 3 days after catheter was removed. Remarried three years after treatment, answers are for 2nd wife. Currently widowed with no partner, still have spontaneous erections.

107	Difficult to keep erections for penetrative sex. Able to masturbate to orgasm.
108	Did penal injections
109	did not have a partner until 2 years after treatment
110	Did not get spontaneous morning erections until the past few months. Never get full erections the way I used to (neither as hard nor as spontaneous)
111	Devoloped ED issues approx. 6 months ago.
112	Decline in firmness or duration, but not in ability to have an erection.
113	Currently 7 mos. post RRP, erections only with Bimix injections. No return of nocturnal tumescense
114	cannot achieve an erection after my surgery.
115	But implant is total success - any time, for however long I want.
116	Both nerves taken during surgery...
117	Between RPP {non-nerve sparing, slash and burn} and Androgen Deprivation Therapy I have not had a spontaneous erection, or any other kind
118	Began taking Cialis 3 months post-op and things improved rapidly on the ability to get an erection
119	Before radiation had erection semi-hard, after radiation no erection.
120	Bad question, Very unhappy married life. I probably could if I were in a different cituation.
121	answer to above 23, 24 & 25. had problems with erection before bding diagnose; however used a vacuum pump
122	An erection requires some stimulation and at the year point, isn't as firm as prior to surgery, but enough to make penetration.
123	All erections were partial.
124	All erections have to be manipulated.
125	Again, my only treatment has been active surveillance, so this question probably doesn't really apply to me.
126	After surgery didn't work.
127	after surgery did penis was shrunk
128	After nine months the penis shows some increase in size when stimulated.
129	8 years after treatment, my libido began to lessen
130	50% of time use 25mg-50mg Viagra
131	4 months post RP now
132	12 years

The range of answers and emotions expressed is a tribute to all the men who participated in this survey. The information contained in the above comments would endorse and support current thinking and recommendations dealing with the importance of exercising penile tissue after prostatectomies. Low doses of oral PDE5i medication, Viagra® or Levitra®, should be started two weeks before treatment. Two weeks after the catheter is removed, full doses of Viagra or Levitra are started once per week. On the remaining six nights, low doses of Viagra or Levitra are prescribed. At four to six weeks after treatment, penile shots (Bimix or Trimix) are recommended to achieve 60% of a full erection two to three times per week. On the non penile injection days, low doses of Viagra or Levitra are suggested. A penile pump may be substituted for the shots.(1.2.3)

26. What things did you try to regain an erection? And in what order?

Answer Options	Tried 1st	Tried 2nd	Tried 3rd	Tried 4th	Tried 5th	Tried 6th	Tried 7th	Response Count
Nothing	86	5	3	0	1	0	0	95
Self masturbation	137	62	24	14	4	1	1	243
Mutual masturbation	31	59	40	10	8	3	1	152
Oral sex	8	38	45	27	9	3	1	131
Sex toys such as	3	1	7	12	11	4	2	40
Sex therapist	2	2	1	0	1	1	1	8
Erotic magazines	3	9	10	5	4	0	0	31
Erotic movies	9	16	21	21	11	5	5	88
Penile shots	11	15	20	19	20	16	5	106
Pump	19	40	29	23	18	8	2	139
Urethral	2	3	9	9	5	2	2	32
Penile implant	4	3	5	4	4	2	5	27
Oral medication	97	64	54	41	9	12	2	279
Other	3	3	3	1	1	0	1	12
Oral medication name								285
						answered question		426
						skipped question		59

Self masturbation and oral medications were most frequently used to obtain an erection, but several other options were tried. This needs to be evaluated by treatment and age.

27. How important do you think is penetrative sex is?

Answer Options	Not important	Not very important	Somewhat important	Important	Very Important	Response Count
For you	25	32	75	161	146	439
Your partner	48	68	90	108	74	388
					answered question	441
					skipped question	44

The partners were more upset by lack of penetrative sex than the survivors thought they were. When the partners were matched, however, the partners were less upset than their survivor partner guessed. (See Partner Q13)

45

28. Would you have a penile implant?

Answer Options	Response Percent	Response Count
Yes	9.9%	43
No	57.9%	252
Maybe	32.2%	140
	answered question	435
	skipped question	50

This data needs to be analyzed by treatment and by age.

29. If you were to have a penile implant, who is it for?

Answer Options	Response Percent	Response Count
You	22.7%	65
Your partner	13.3%	38
Both of you	64.0%	183
	answered question	286
	skipped question	199

Most men who had penile implants were satisfied with the results.

How important is it to experience an orgasm on a regular basis even if it does not result from penetrative sex?

Answer Options	Not important	Not very important	Somewhat Important	Important	Very Important	Response Count
For you?	35	33	108	134	125	435
How do you think your partner would answer?	59	84	99	112	48	402
Comment						38
					answered question	435
					skipped question	50

Q30. Respondent Comments

1	Wife on bp an depression meds, suppresses libido
2	Wife is totally non-interested in sex. Feels people our age should not worry about ED and other sexual issues and just be glad of the successful cancer treatments.
3	Wife hates sex - we have not had sex for > 15 years
4	We both enjoy making love very much, even after 44 years
5	That's what partner has said, and how she behaves
6	Spouse is very supportive...she would be there for me whether or not I performed at all...but we both still enjoy sex and wish to preserve it as long as possible...
7	spouse has lost a lot of libido post-menopause
8	So far, I always have an orgasm.

9	So far so good
10	Since her hysterectomy she has lost interest in intercourse.
11	She's 67 and not real worried about it
12	She does not think she has ever had an orgasm.
13	Post treatment orgasms VERY difficult to experience during intercourse
14	partner had multipule orgasm w/penetration or orally
15	Partner doesn't have orgasms
16	our sex has been infrequent since dx and tx
17	not very important
18	not really too sure anymore...
19	No longer have a normal orgasm (dry)
20	No current partner
21	my wife passed away 2 years ago
22	My wife has always been non-orgasmic
23	My wife accieves orgasim with use of vibrator
24	My partner thinks it is important for me to have an orgasm and does her best to achieve that
25	It's very important to my partner for her to have an orgasm. I believ it's also important to her for me to have an orgasim.
26	It has been over ten years. Come on now. Maybe you need someone younger for this survey.
27	I think wife answer would be because my inability
28	I take Citalopram as an anti-depressant not associated with my surgery. However, it does make it more difficult to reach orgasm.
29	I probably get the urge once per week.
30	I do satisify her manually.
31	I cannot have as orgam since the operation, it would be dry.
32	He as issues based on blood pressure meds.
33	Diagnsis at 65 jears
34	Bad question.
35	An erection is not needed for this - but life w/o orgasm - no way!
36	AM TOLD PARTNER HAS LOW LIBIDO.
37	Although I have always been out to her, I only have sex with my wife and have always loved sex
38	Again, not a real parter but friend on several years.

The bar graph at the beginning of this document shows that 60% of the survivors and 40% of the partners felt that having an orgasm was important. Only 8% of the respondents felt it was unimportant. Again, this needs to be analyzed by age and treatment.

Although only 8% of the survivors included comments, they made some important observations.

31. How often do you have penetrative sex now without any form of assistive aids: pills, pump, penile shots, urethral suppository oe penile implant?

Answer Options	Response Percent	Response Count
Not at all	75.6%	334
Less than once a week	16.7%	74
1 or 2 times a week	6.6%	29
3 times a week or more	1.1%	5
answered question		442
skipped question		43

Almost 98% of survivors answered this question. A large number, 75.6%, were unable to have penetrative sex without some form of assistance. It is important to remember that this includes all forms of treatment. Only 50% of respondents had prostatectomies. Do men overstate their ability to have penetrative sex? This data needs to be analyzed by treatment type and age.

32. How often do you have penetrative sex now using assistive aids: pills, pump, penile shots, urethral suppository oe penile implant?

Answer Options	Response Percent	Response Count
Not at all	62.2%	267
Less than once a week	25.6%	110
1 or 2 times a week	9.6%	41
3 times a week or more	2.6%	11
answered question		429
skipped question		56

32. How often do you have penetrative sex now using assistive aids: pills, pump, penile shots, urethral suppository oe penile implant?

Answer Options	Response Percent	Response Count
Not at all	62.2%	267
Less than once a week	25.6%	110
1 or 2 times a week	9.6%	41
3 times a week or more	2.6%	11
answered question		429
skipped question		56

32. How often do you have penetrative sex now using assistive aids: pills, pump, penile shots, urethral suppository oe penile implant?

Answer Options	Response Percent	Response Count
Not at all	62.2%	267
Less than once a week	25.6%	110
1 or 2 times a week	9.6%	41
3 times a week or more	2.6%	11
answered question		429
skipped question		56

With assistance, the ability to have penetrative sex increased overall by about 14%.

The biggest increase was 25.6% in the "less than once a week" group.   Again, the data needs to be analyzed by treatment and by age.

33. Which of these forms of assistance resulted in an orgasm with or without penetration?

Answer Options	Tried	Did not achieve orgasm	Achieved orgasm infrequently	Achieved orgasm most of the time	Achieved orgasm always	Response Count
Nothing	60	24	11	13	12	99
Self masturbation	92	28	50	140	108	334
Mutual masturbation	52	21	36	83	49	201
Oral sex	53	34	28	56	43	170
Sex toys such as vibrators	22	8	9	16	9	50
Pénile shots	32	13	11	31	29	89
Pump	50	33	24	28	12	113
Urethral suppositories	17	9	5	4	2	28
Penile implant	10	3	4	8	6	25
Oral medication (provide medication name in comment box below, if possible)	94	52	34	51	41	203
Other	3	1	3	2	0	8
Oral medication name and/or comment						182
answered question						421
skipped question						64

94% of the respondents answered this question. Evaluation of the success of various methods of achieving an orgasm revealed that most of the time or always, self masturbation headed the

49

options at 74% (334 responses), followed by mutual masturbation at 66% (201 responses), oral sex at 58% (170 responses) and oral medication at 45% (203 responses). This raises questions about the efficacy of oral medication.

Are these choices age-related or treatment-related? Analysis by treatment and age would be useful. Does this vary by country?

34. If you have tried alternative means to reach an orgasm, was it for?

Answer Options	Response Percent	Response Count
You	40.4%	124
Your partner	6.2%	19
Both of you	53.4%	164
answered question		307
skipped question		178

Is it more important for the survivor, partner, or both to reach an orgasm?

35. Can a man have an orgasm without ejaculating?

Answer Options	Response Percent	Response Count
No	1.6%	7
Yes	90.6%	397
Not sure	7.8%	34
answered question		438
skipped question		47

36. Can a man have an orgasm without an erection?

Answer Options	Response Percent	Response Count
No	9.6%	42
Yes	69.5%	303
Not sure	20.9%	91
answered question		436
skipped question		49

Orgasms without an erection or ejaculation are possible. Dry orgasms are normal after prostatectomy. Out of 97% of survivors who answered this question, 20.9% were not sure about the possibility of having an orgasm without an erection.

37. Would you agree that women get a great deal of security and sexual pleasure from pure intimacy, kissing, hugging, holding hands, lying close together?

Answer Options	Response Percent	Response Count
No	2.3%	10
Yes	89.2%	389
Not sure	8.5%	37
answered question		436
skipped question		49

A significant number of survivors feel that women get a great deal of security and sexual pleasure from pure intimacy, kissing, hugging, holding hands or lying close together. Survivors should remember that even if penetrative sex is not possible, maintaining this type of intimacy is extremely important. Do not neglect this!

38. What percentage of women do you think experience vaginal orgasm?

Answer Options	Response Percent	Response Count
0%-10%	7.1%	29
11%-20%	15.9%	65
21%-30%	22.8%	93
31%-50%	28.9%	118
51%-70%	16.4%	67
71%-90%	7.1%	29
91%-100%	1.7%	7
answered question		408
skipped question		77

Experts state that only 24% of women experience vaginal orgasm.

39. Did you both go to the appointments with your urologist?

Answer Options	Response Percent	Response Count
No	27.2%	117
Yes	51.4%	221
Occasionally	21.4%	92
	answered question	430
	skipped question	55

Two sets of ears are better than one. Treatment for prostate cancer affects both partners so it is a good idea for partners to go together to appointments and for both partners to pose questions.

40. Did your treating physician discuss any of the following?

Answer Options	Yes	No	Not sure	Response Count
That after treatment male sexual function would be different requiring some adjustments.	323	98	12	433
That women need to be wanted and needed physically and mentally.	48	333	31	412
That for women foreplay provides closeness and intimacy.	51	340	19	410
That orgasms are important for women.	22	358	28	408
That few women have vaginal orgasms.	7	375	24	406
That clitoral orgasms are stronger than vaginal orgasms.	15	376	18	409
That manliness and masculinity are closely tied to having an erection and that this issue is manageable.	61	332	20	413
Libido remains the same even if sexual function does not or partially returns.	88	301	27	416
Male orgasms will be dry. (no ejaculation)	288	131	9	428
That possibly penetrative sex may not end in an orgasm.	92	292	32	416
That it is important that partners maintain intimacy, hugging, kissing, caressing, lying naked together.	81	316	18	415
How often you had sex.	103	292	18	413
My physician discussed erectile dysfunction issues.	318	99	12	429
			answered question	439
			skipped question	46

Some physicians treating this group of survivors were extremely savvy when it involved informing their patients about changes in sexual function. Dealing with the other issues was poor.

It would be interesting to know whether the patient asked the questions to elicit answers or if the treating physician initiated the discussions.

41. If your physician discussed with you the possibilities of having problems with sexual function after treatment, was there a percentage of problem possibility provided? MUST BE WHOLE NUMBER NO DECIMALS!

Answer Options	Response Average	Response Total	Response Count
Percentage	44.02	9,112	207
		answered question	207
		skipped question	278

Answers ranged from less than 10% to 100% so your guess is as good as ours! It would seem to depend on how comfortable the treating physician is with discussing sexuality issues.

42. Overall how big a problem do you consider your sexual dysfunction to be for you?

Answer Options	Response Percent	Response Count
No problem	9.4%	41
Very small problem	11.4%	50
Small problem	16.7%	73
Moderate problem	38.4%	168
Big problem	24.0%	105
answered question		437
skipped question		48

For 62.4% of the survivors, erectile dysfunction was a moderate to big problem. This data needs to be analyzed by age and treatment.

43. If you are on androgen deprivation (in the form of shots or similar treatments to depress or eliminate your testosterone) are you able to have an orgasm with a:

Answer Options	Response Percent	Response Count
Normal erection	2.5%	8
Assisted erection	4.3%	14
No erection	13.7%	44
Does not apply	79.5%	256
answered question		322
skipped question		163

14.7% of the survivors who responded had or are having treatment to suppress or eliminate their testosterone, and 66.7% are unable to have an orgasm. This finding needs to be analyzed by age and treatment.

44. How big a problem do you think the sexual dysfunction is for your partner?

Answer Options	Response Percent	Response Count
No problem	17.4%	71
Very small problem	22.5%	92
Small problem	26.5%	108
Moderate problem	25.5%	104
Big problem	8.1%	33
answered question		408
skipped question		77

There was agreement when these answers were compared to those of the partners, except when it concerned the "big problem." The partners were much more upset than the survivors thought. This data, however, came from unmatched partners. When the partners were matched, the partners were less upset than the survivor guessed. This could be due to the matched partners being together for many years. Further analysis of the data of the matched partners would be helpful.

45. Where did you obtain your information on prostate cancer?

Answer Options	Response Percent	Response Count
Library	19.6%	86
Magazine	17.6%	77
Book	51.1%	224
Internet	85.2%	373
Cancer support group	54.1%	237
Friend	16.2%	71
Prostate cancer survivor	42.9%	188
Physician	74.4%	326
Other	4.6%	20
Other (please specify)		42
answered question		438
skipped question		47

Q45. Respondent Comments to "Other (please specify)"

1	Was a cancer researcher for 35 years. Knowledge from research collaborations in the medical community.
2	Very disappointed in physicians ability to provide information
3	UsToo Website
4	US T0) organization
5	Two Drs. really...Radiologist as well as my urologist
6	Treatinig facility.
7	Throughout the whole process I did very extensive research mostly through the internet, focusing especially on peer reviewed puplished studies (and .edu and .org web sites).
8	The Wellness Center - (very good and supportive!)
9	Specialist (surgeon)
10	SECOND OPINIONS - SURGICAL UROLOGIST AS WELL AS PATHOLOGIST.
11	Radio at times, meetings and weekly telefone program.
12	Prostate clinic
13	Prostate Center VGH
14	prostate cancer support nurse
15	prostate cancer research
16	physicians dicussed mainly treatments. No mention of active surveiliance
17	PCP
18	PC-Forum
19	PCAI forum (Prostate Cancer and Intimacy)
20	patient treatment experience

21	Online cancer forum
22	on line, Johns Hopkins, Mayo
23	National cancer seminars
24	My wife is a medical proffessional. She works closely with Urologists.
25	My original urologist could care less about impotence. He insinuated I had failed him for not recovering potency. He hated for my wife to be present. Was very unsupportive. The urologist that performed my penile implant is ABSOLUTELY WONDERFUL !!!!
26	My doctor did not mention support group, which I think is very important. I had to find local support group on my own, after surgery
27	My answer to 44 is because my implant is a total success.
28	Multiple interviews with leading physician specialists
29	Lecture given at local hospital
30	Journals
31	John Hopkins Pamplets
32	Internet
33	I researched it throughly.
34	I did a lot of reading
35	I am a physician.
36	I am a physician
37	Father
38	Doctor Strum P2P
39	Diagnosed at The XXXXXX Clinic (XXXXXX) Time between diagnosis and surgery; 9 days. They were very supportive and supplied a lot of information. I spent many hours on-line researching.
40	Cpri
41	anywhere I can find it.......
42	American cancer society

Most survivors obtained information first from the Internet, then from their physician, cancer support groups, books and other prostate cancer survivors.

Only 9% of the survivors commented on where they obtained their information.

46. Support Groups If the answer to "a." is "No" go to "b." If the answer to "b." is "No" go to "c." If the answer to "c." is "No" go to the next question.

Answer Options	No	Infrequently	Often	Very often	Response Count
a. Did you attend a support group prior to treatment?	329	52	26	27	434
Did your partner go with you to this group?	142	31	19	11	203
Did this group help you?	79	21	45	35	180
b. Did you attend a support group after treatment?	183	79	80	83	425
Does your partner go with you to this group?	172	51	32	25	280
Did this group help you?	69	49	91	57	266
c. Do you attend a support group now?	250	57	43	71	421
Does your partner go with you to this group?	175	37	18	21	251
Did this group help you?	85	37	69	53	244
				answered question	442
				skipped question	43

Prior to treatment, 12% of men attended support groups; after treatment, 36% of men attended, but only 28% of the partners.

47. If you attended a support group, kindly answer the following?

Answer Options	Response Percent	Response Count
Could you describe how the support group helped or if it didn't, why it did not help?	97.9%	228
Could you add any suggestions for improvement of the support group?	57.1%	133
answered question		233
skipped question		252

Q47 part 1: "Could you describe how the support group helped or if it didn't, why it did not help?" Respondent Comments

1	You find others with similar problems. That is how I found out about artificial sphinctors.
2	Yes, helped. Men with similar experiences willing to discuss it. Not alone.

3	Weekly speakers give information on new drugs, alternative therapies, nutrtion,new treatments, holistic and naturopathic remedies techniques etc	
4	We have very good speakers and programs every month. I have developed solid friendships there.	
5	We could and did talk about issues nobody's doctor seemed to talk about, such as the loss of sexual function and how to deal with that emotionally.	
6	We bring in speakers on various subjects which are great	
7	Was of great help after I decided to go about 1 week after my surgery.	
8	US Too in the Chicagoland area was a terrific support both before and after my surgery.	
9	Understood that my somewhat limited ability to have an erection was better than many that had NO ability following surgery.	
10	TYPE OF FOODS TO EAT OR AVOID, ETC	
11	Totally helped talking to others with PC and sharing informaton. Very fortunate to have two extremely knowledgeable PC survivors of 17 years each who have researched PC and are willing to share their infomation and feelings.Listening to them and others has proven at tims to be better than listening to the Doctors in the various fields of treatment, as that comes off as an informercial.	
12	too many guys delaying treatment, looking for the "perfect" treatment that doesn't exist	
13	too general/ not supported by local medical field	
14	Time was an issue	
15	thru regular communication via email	
16	They were not interisted in the sex part of prostate cancer	
17	They are not concerned about sex as much as me only staying alive	
18	The UsToo support group was attended primarily by older men. It didn't focus on the sexuality issues.	
19	The support group talked about the different treatment options, gave literature and additional resources.	
20	The sexuality problem not discussd	
21	The men were a lot older than me. Sex was not a topic of discussion.	
22	The group helped by honestly answering questions	
23	talking w/ survivors as to what to expect and how to cope w/ it. good old caring fellowship!	
24	Talk to other people that have gone thru Prostrate Cancer	

25	Support group helped understanding various treatments, effects, experiences and living examples; assess doctors and hospitals.
26	Supplied a library of info. Group discussions with other men. Gave you hope and courage when needed.
27	speakers came and really helped on impotence and incontinence...couples discussed..
28	Since my cancer was detected early, my situation was not as complex as some of the other guys'.
29	showed me ws not alone; brought up different treatment modalities; support groups, per se, did not provide much clinical information; other doctors werre mentioned; REAL Help was emotional, and carthartic
30	Showed me that this cancer was survivable
31	Showed me that other cancer survivors existed and had lives
32	Sharing of similar experiences
33	sharing of information
34	sharing information
35	shared resources
36	Shared our experiences, an indication of what you could possibly expect. what Q's to ask your Dr.
37	Shared experiences, sources of information, encouragement
38	Shared experiences with treatment and recovery stages.
39	Same problem to discuss
40	References, answers and personal experiences
41	recognize that everyone is/had differing situations
42	Put my choice of treatment in perspective. It was helpful to know my problems were not unique and moderate in comparison to some others.
43	Providing alternatives to consider to deal the various problems
44	Provides others who can relate to what you're going through; Provides updated information on cancer survivor recommendations; Able to hear others story or experience to compare to mine
45	Provider and treatment information, sharing of experience with other patients.
46	provided support & info on PC to a survivor
47	provided information about PC
48	Provided info on prostate survival, also showed me many individuals who were surviving the cancer
49	Provided info from oter survivors
50	Provided info and support listening to others experience and to guest professionals

51	Provided info (or references to info) about tests, treatments, side effects, doctors, and diet.
52	provided first hand information
53	Provided bags of information - too much really. Gave me perspective on initial difficulties with proposed treatments, which helped me decide treatment. Later provided contact with other survivors who had similar urinary blockage problems. Where it didn't help was it was difficult for me to take part in general discussions due to poor hearing. Obviously there was little they could do about that, they did what they could, I have no complaints on that score.
54	provided a lot more info than doctor
55	Prostate group was useless - men were to uptight. XXXXXXX Cancer support group was very helpful
56	Prostate cancer and other medical education and the opportunity to help others
57	practical, first hand experiences, the ones the doctors never mentioned
58	Positive for helping others; Negative for mental state-now choosing to not focus on cancer and negatives.
59	Poorly organized. No attempt to relate experiences. Meetings in a church - "fear" of dicussing sex in those environs.
60	Pertainent info., also where to find it
61	personal experiences helped, how people deal with issues talks by professionals gives updates
62	people with like problems sharing and helping to educate newly diagnosed individuals
63	People could not relate to my problems as they had been in the support group for a long time, like 15 years.
64	Patients' participation ineffective.
65	Out of appx. 20 men, I am the only one who has had a radical operation.All the others have had radiation and 1 was by laser treatment.
66	Opened discussions about things that I am experiencing
67	open talk on many issues, guest speakers very good
68	Open dicussion of medical and sexual issues
69	only went once no discussion on topic
70	Only some - being with men and women who are coping helped. Their generousity helped. Their reluctance to let us talk openly greatly hindered and resulted in my dropping out.
71	not enough interest in this area
72	No one else seemed to have the same problems I was having.
73	no new information

74	new information and answer questions
75	New information
76	N/A
77	My circumstances were well defined and I didn't need the group dynamic.
78	Most of their time seemed to be directed to people who wern't there, pre-cancer subjects
79	Most of the survivors were of the radical prosectomy procedure and the Dr. that volunteered to attend the meetings was a surgeon who advocated this procedure, so we felt as outsiders
80	Most discussions seem to be oriented towards attendees who are still battling prostate cancer, not issues for those who are in remission
81	Most are advanced cancer survivors, not expectant management
82	more focused on publicity and fund-raising than mutual support, all meetings are large group with no openings for supportive small group meetings
83	moral support, ideas how to seal with side effects of radical prostatectomy
84	Misery likes company! It is not so depressing when you know that there are so many worse off than you.
85	Mine was a very ealry stage cancer. Most members of the group were incurable cancers. Their advice was not at all appropriate for me.
86	Met other survivors (and learned that people survive). Learned about technical details of prostate cancer and followup treatment as necessary. Learned about incontinence and how to manage it. Learned about diet. Learned about Vitamin D3. Received encouragement.
87	Men discussed effects on sexuality and provided common understanding. Dr. lectures informative
88	Many speakers cover what to do to prevent Prostate Cancer or new treatment techniques.
89	Makes me realize I'm not alone
90	Made us not feel like we were alone.
91	Loma Linda support group was great. At home it was such a depressing experience I decided I did not need it in my life
92	Listening to the storys of other men and their recovery process.
93	links to information sources
94	let me talk to others who have or are experinceing the same thinks
95	Just talking & Listening

96	Just nice to know all the types of real people that it affected + how they handled it - in person
97	just knowing others are in the same boat so to speak
98	Just didn't address some of my problems and I did feel comfortable in talking about some of my problems.
99	It was more of a medical lecture than a support group. I'm woring with the facilitator now to change the format of the group.
100	It was great for post-diagnosis to discuss treatment options. It is lacking in terms of support for ongoing adjustment issues as it really is about the newly diagnosed patients.
101	It was dominated by one man whose sole mantra was to get cox inhibitors and rage against eating meat.
102	It was depressing.
103	It provided alot of answers for me and now I can support new people with that.
104	It made me more curious about the surgeon I selected, I wanted to find a surgeon who was sensitive to my sexual concerns.
105	IT MADE ME AWARE OF MY CONDITION. I THINK I BLOCKED THINGS OUT. I DID NOT WANT TO BELEIVE.
106	it helps by showing that there is a good QOL after prostate cancer
107	It helps by seeing how others cope with the same problem and to learn more about other problems
108	It helps because the group is an opportunity to compare symptoms, share experiences, learn latest PCa news.
109	It helped put my problems into perspective.
110	It helped in that I saw men how had survied PC and life goes on. Meeting are all geared around first time attendees
111	It helped in putting things in prospective. I was no alone in this fight. Others had it much more severe than I. This was a support group for all forms of cancer. It helped to just talk it through with others in the same fight I was.
112	It gave me a place to vent to other patients.
113	It did not help me. I was the youngest person there (50), and it seemed like a function for people that were in much worse shape than I was. I actually felt guilty. I was already having erections and penetrative sex (about 5 weeks after surgery) and many of these men were still fighting the disease. I only went once.

114	Initially the group helped and answered questions about treatments, recovery, etc. However, the group didn't adresss post surgery ED - there was a planned session to talk with a knowedgeable person about sexual aids for post sugery ED but it was cancelled because of bad winter road conditions and it hasn't been rescheduled.	
115	Information shared, speakers gave updates on medical treatment, outcomes, etc.	
116	Information and I help run it.	
117	INFORMATION AND COMPARING PROBLEMS	
118	information	
119	Info mainly about how to treat prostate cancer (PC)...this of little value to me, since have no prostate. Keeps me up on latest bladder incontinence and ED treatments.	
120	I'm still just getting involved. It won't help my sexual probelms. Few men have a 4" penis.	
121	IM A SUPORT GROUP LEADER	
122	I went to local in-person groups and on-line groups both let me know I wasn't alone and gave valuable practical advise on issues from catheter pain to using shots to living my life and interacting with others.	
123	I was one of founders of a XXXX XXX XXXX support group about a year after treatment. Just talking with other men was helpful to me. I wish the group was still functioning, but we stopped having meetings because of very low attendance.	
124	I was much younger than other attendees and the group did not cover full incontinence issues.	
125	I was extremely angry due to impotence. My support group was essential to my recovery. Reaching out to newly Dx'd preserved my sanity.	
126	I was exposed to other men's and women's issues regarding ED and was informed about new literature on the subject.	
127	I was able to find out how men fared with different TX's.	
128	I was able to discuss incontinence,reason I wanted the AIS (AMS800)	
129	I started a support group and worked to add members, then the hospital stopped support	
130	I started the group since there was none operating in my city at the time I was diagnosed. I believe it helps to know that I am not the only one like this.	
131	I run the 3 support groups	
132	I realized that what I was experiencing was normal for my treatment and that I'm not alone.	

133	I only went twice and the subject matter is usually on treatments but never on active suurveiliance
134	I have not had much in the way of informal support from friends or family. The groups that I have been in are important
135	I have acheived more from a one on one with survivors outside a group mainly due to my moving around for employment.
136	I had already learned as much as anyone there.
137	I go to a general cancer support group that is somewhat helpful because of camaraderie and a sense we're in it together. I tried a prostate cancer specific group, but they were MUCH older than me and concerned with advanced cancer (i.e. metastatic) issues
138	I gained valuable insight. But mostly I found it rewarding to help other men.
139	I gained a lot of knowledge about CaP. I was helped emotionally and I helped others. I helped facilitate a group.
140	I found that I wasn't alone, and that non penetrative sex was still possible and still good too
141	I found importants informations in my Forum of Discussion KISP (www.prostatakrebse.de)
142	I facilitate a PC Support Group. It is good for men and their partners to share their experineces, both medical and emotional, and to discover that they are not alone and that what they feel is common.
143	I facilitate 2 support groups. Wish I had known about them prior to treatment. They do help! Great information
144	I experienced the comraderie of fellow men who had the same cancer as I. Though many were much older than I, there was a common bond.
145	I don't need it. I have no depression
146	I do not think a support group will help me with my wife
147	I currently run the support group fot the Island
148	how to cope & what to expect
149	History of other survivors
150	Helps to find out what problems other people are experiencing and how they cope with them. Helps to know you are not alone.
151	helps because other survivors share info about their experiences, they understand what I'm experiencing
152	HELPED: EMOTIONAL SUPPORT; COMMONALITY.
153	helped with details of the process and solutions for dysfunction
154	Helped when relevant
155	helped understand post-op treatment for incontinence issues

156	HELPED TO REMOVE THE FEAR ASPECT OF THE TREATMENT
157	Helped to provide information athat I did not research on my own; information was from health professionals as well as other survivors
158	Helped to know that I was not alone - that other men were dealing with the same problems. Helpful to hear how they were dealing with the same problems and issues.
159	helped my psychologically
160	Helped me in determiningtype of treatment
161	Helped me determine the treatment options I wanted to avoid.
162	Helped by discussing what we all go through, and supporting each other, and offering research material
163	Helped at first. Small group. Same stories at every meeting got boring. Group disbanded for lack of interest.
164	help me in that I am learning to be a support group leader, which in turn I become more educated in regards to PC
165	Hearing others talk openly about their problems and what they have tried to do
166	hearing how other men handled their pca
167	Having others with shared experience and being able to talk openly.
168	Hasn't helped since I was basically over the emotional part of recovery before I started going.
169	Group provided comfort and support, answered my questions, urologist speaks to us and answers questions
170	group described individual symptoms
171	Great speakers who address problems common to all of us.
172	Gp. members talk about their Ca and their treatment openly without demeaning others for their decision.
173	Good to speak to others who were going through the same thing
174	Good to speak openly with others who understand and have similar experiences
175	Good speakers brought in and people willing to share and answer questions
176	Good information, but after the fact!
177	Good information from folks who've been or are there
178	good info and insights from others. frequent updates on pc
179	Good Info
180	gave me new information on treatments, survivors experiences, and support I did not know was available before treatment
181	Gave me an opportunity to talk to other survivors

182	gave me advise on how to proceed after operation when discovered cancer was outside prostate..
183	Gave me a pathwaty to others who had been diagnosed and what kind of treatmnets were done, how side effects affected them.
184	gained over knowledge from speakers
185	Gained knowledge from others and also physicians attending
186	Gain from others experiences.
187	Friends
188	Frank discussions on all topics
189	Found surgeon from second opinion
190	For the most part, nothing new was discussed.
191	Focus is toward older men. We have a female nurse who facilitates our group. She is a wonderful person, but her presence is chilling on frank and open discussion
192	Fellowship and visiting Medical Doctors alternative opinions
193	Fellow prostate cancer survivors -- wonderfully helpful perspectives!
194	Experiences and new therapies
195	exchanging experience is always helpful, especially when you havve little experience on the subject.
196	EVERYONE IS DIFFERENT, WHAT WAS RIGHT FOR ONE MAY NOT BE FOR ME
197	everyone is different, others had different treatment, lower PSA
198	Encouragement through shared experiences. I just started and have only been once
199	EDUCATIONAL AND SHARING OF SURVIVORS IN EACH JOURNEY
200	Dose not help on a specific case. It is very generic and about new products or treatments
201	discussions and medical members attending
202	discussion on mutual problems helpful, as well as informed speakers
203	Discussing new forms of treatment and side effects.
204	Discuss the problems of therapy and care and which provider offers the best care for PCa survivors
205	Discuss "real world" problems with prostate cancer survivors
206	different problems facing prostate cancer patients
207	Didn't really feel that I needed support
208	CSACASCAS
209	could not attend due to distance & sinus lining intollerance to poluted air - bubble boy symptoms-
210	Could identify with other men of similar experience

211	comfort in learning that I am not alone in my experiences, feelings, sense of loss
212	Clarifying issues not discussed by treating physician
213	Cancer doctor speakers always provide some new information.
214	can talk openly, all in the same boat, knowing others are feeling the same and going thru the same thing and it helps
215	Better understanding of problems; assistance to others
216	Better education overall and great understanding of concerns
217	Being bisexual and having additional medical problems with RA, as well as deceased spouse and children I raised after their mothers' death, my daughter who lives with me just turned 22 at diagnosis and I seemed pretty much "closeted" and presented rather unique psychological problems.
218	Being able to share and get advice
219	attended 2 different groups for many years. got to hear multiple first hand experiences of a wide variety of treatments. was exposed to excellent presentations on many post treatment issues, and how to deal with them.
220	At this time my cancer is in remission. ED is not one of our frequent topics. I am one of the leaders of our support group. We work mostly with newly diagnosed or those with recurring PC problems
221	Answered questions
222	Allows for the sharing of concerns and ideas provides professional updates to PCa
223	age difference
224	After treatment, I had many problems that we could discuss. This led me to see that they were not uncommon and that they would hopefully be resolved with time.
225	ACS group, generic comments
226	ability to talk with men who underwent a similar experience
227	Ability to talk to survivors
228	A wealth of information, there is so much someone picks up what you miss.

Q47 part 2: "Could you add any suggestions for improvement of the support group?" Respondent Comments

1	yes-discussions on what to do if an operation that had complications leaves a patient such as me in a recovery situation of 2 years and is still ongoing.
2	Yes

3	Yes
4	Would be nice to share some more personal things but group does not know each other well. I am in a Mens Group at my church and they are much more supportive emotionally.
5	Wider publicity
6	We need effective and informative programs; we have some difficulty with this now.
7	We have a very active group, probably on of the best in the nation
8	We have a great Group --70 at each meeting ...no wives present
9	UsToo is a great organization - I had no problems with them
10	Use e-mail for info of meeting time, LOCATION, and future speakers.
11	Too early to say.
12	They are doing a great job
13	The group needs to be facilitated by persons who know how to remove reluctance to discuss sexual matters openly.
14	The group is small. It's rightly focusing on prostate cancer treatment options. We are looking to improve outreach to the community.
15	the doctors seem afraid of the group and they need to support it more
16	Talk more about ED after surgery
17	Supports groups should have a positive side to the meetings. Just listening to problems is depressing.
18	Support groups in general need more diversity and community support (volunteer and financial)
19	Support groups are good. Problem is publicity, exploration of treatments for newly diagnosed.
20	Stop the business side and get more on the medical side in layman terms.
21	STAYED INVOLVED AFTER TREATMENT OPTION.
22	Specific treatments, successes, failures, recommendations
23	Sometimes a little negative and probably too focused on the cancer returning in one form or another
24	Some members are insensitive to new members fears
25	shared experiences
26	Sex needs to be a topic for descussion.
27	See above.......40-60 year old sexually active men get prostate cancer too.

28	REAL information about treatment modalities is needed in unbiased presentation along with side effects, consideration of a lifetime of chronic diesase; A unified general information pack could be developed by organizations to give to newbies so they could start with an overview and hope.
29	questions and open answers most useful
30	Probably more important to attend PRIOR to deciding treatment.
31	Proactive guidance to group dynamics - allowing more than one person to talk - many small groups instead of one large - more emphasis on individuals talking and less on health professional speakers
32	POST-OPERATION INFORMATION, ETC
33	Physicians, specialists must contribute with presence, lectures.
34	Perhaps there should be different groups for: a)still in treatment and/or untreatable, b) have finished treatment and need support for this stage, and c) I have regained whatever sexual function I will and how to cope.
35	Perhaps more discussion of intimacy issues--an expert to help guide the discussion would be good.
36	Online
37	One support group was open to spouses, another one was not so we don't participate in that one anymore but we believe it is important for spouses to be included
38	one lady talked too much
39	not to be afraid to discuss ED and the truth after surgery - 60-80% will have some sort of problem.
40	Not sure. I'm not all that social, so there has to be a lot of learning to draw me, and after a couple of years that was no longer the case.
41	Not really ... if my PSA goes up again and further treatment is needed, I will probably return.
42	Not at this time, we are almost overloaded with our current program
43	not at this time
44	NOT AT THIS TIME
45	no suggestions offered on ways of having intercourse; I did attend a presentation on penile implants nut cost and caveats were turn-offs.
46	no longer remember well enough
47	no as it was a small community and it was over an hour drive for me to go to a larger community
48	No

49	No
50	No
51	No
52	No
53	No
54	No
55	No
56	No
57	No
58	No
59	Needs better leadership and control of meeting
60	need to ger awareness out to public
61	Need to find a way to attract people with "good" stories to tell. Seems to be primarily a place for people with problems to look for answers and simpathy.
62	Need to belong is less urgent after treatment and non-problematic recovery.
63	need more expressive facilitation, not just sharing resources or general condition at present
64	need more examples and discussions of positive outcomes
65	Need discussion of emotional response to diagnosis, treatment, sexual functioning.
66	N/A
67	My support group, Us TOO!,is tops.
68	more volunteers
69	more time on questions
70	More things that address my age group
71	More support from urologists
72	more support from prostate cancer survivors who are in the public eye.
73	More realistic & inclusive presentation of the treatment options. Proton therapy wasn't mentioned until I asked the qeustion.
74	More professional speakers & people being more honest about their problems.
75	More people participating in the organizing of the meetings
76	More of a focus on EMOTIONAL issues directly tied to erectile dysfunction and the impact on the survivor/partner. The support group dealt minimally - if at all - with emotional effects as opposed to physical effects.
77	More membership. Would be good if more men affected by PC wouild attend and participate.

78	More information of how to please your wife and make sexual intimacy a part of our lives again.
79	more groups addressing various sexual identities, as well as psychological support.
80	More frequent smaller groups
81	more doctors
82	More discussions on sexual disfunction problems including ED, orgasmic dysfunctions and intimacy related issues.
83	More discussions of interest for those without a prostate. Sex, incontintence long term odds , rise PSA treatments
84	Men are jerks in these groups on there own - need woman present.
85	Match the advice, or contact people of the support group to the new attendee's cancer stage. I was continually bombarded with inappropriate (for my stage cancer) advice. Also the group should limit anecdotal advice (which in many cases was untrue in general) and try to stick to factual information. I almost felt the support groups (I went to two) were harmful to me. I am glad that I did not take their advice.
86	male physican, intern, or nurse facilitate group.
87	Local support group was centered on those who were in watchful waiting and were not necessarily supportive of those who took action. Managing balance is important between those who have varing interests in approaches.
88	It is a group. It is my wife causing the issue.
89	Increase the material for those in expectant management
90	increase discussions of ED
91	Improved outreach to encourage general public to do testing
92	Important that PC Survivors include their spouses or significant other in all facets of PC, particularly at visits to doctor as 4 ears are better than two.
93	I went to only one meeting of support group. It was informative.
94	I thought we had a very good group I think it met the needs of all cancer types
95	I think the support groups need to be oriented more toward future treatment decisions of newly diagnosed.
96	I helped form another support group that was not as technical as my other group. Information is at times more social rather than technical.
97	I haven't a clue.
98	I have

99	I enjoy hearing about other men's problems, but I want to help those that have yet gone through the change.
100	I don't think men are open to discussing ED, leaking, etc. with other men
101	I am not sure that having partners there is a good idea.
102	HIGHER LEVEL OF COMMUNITY AWARENESS
103	have speakers on nutrition and other alternative choices
104	have sharing sessions occasionally
105	Have doctors tell their patients ahead of treatment.
106	Have doctors tell patients about these groups BEFORE treatment is decided upon.
107	have a healthy free breakfast available
108	Have a breakout group for those getting used to the new normal so that those three months out can talk to those three years out and have some sense as to how things may continue to improve and coping strategies.
109	group i attended was very good
110	Great support and wealth of information for those in need and continuing problems with there PSA
111	get younger men involved
112	get some Goverment help; financing etc
113	Get more than the usual 4% of diagnosed to attend and the newly diagnosed to attend to learn about this chronic disease.
114	get more info on partner needs intimacy etc..
115	Fine as is
116	filter out discussion of "spurious" treatments
117	emotional support
118	does a good job
119	Doctors should encourage this preceding any treatment
120	Do not have it on Sunday evenings
121	Did not know there were support groups prior and now the damage is done.
122	did not attend as I sis not know where to go, need better awareness of meetings
123	CSCAACASC
124	closer to where I live
125	Closer to home. Need to drive 30 miles for nearest group.
126	Broaden discussion/presentation topics beyond just PC treatment and fundamental incontinence and ED treatments. Emphasize that an active life, including sex, is quite achievable despite damaged plumbing. Emphasize that early PC detection is very important--every male should know this!

127	Both groups were excellent (particularly at XXXX Hospital in XXXXX). NOTE: I have been single since 1992; was alone at diagnosis and there was no partner to attend the groups with me; have had several relationships since treatments and only once did a partner at that time attend with me.
128	Better attendance would help, but hard to make it happen.
129	better attendance
130	be more open about sexual issues, esp. for younger men!
131	Attendees should stop trying to force beliefs to those that the recommendations were no longer options
132	Ask each that share their story or experience to relate similiar information such as when diagnosted, PSA levels at certain time, their family medical history as it relates to prostate cancer, etc.
133	addressing more needs of incontinence and issues for gay patients

30% of survivors made suggestions on how to improve support groups.

Most respondents wanted more information dealing with erectile dysfunction as well as more volunteers, publicity, professional speakers and physician support.

48. Do you make use of the internet as a support network?

Answer Options	Response Percent	Response Count
No	29.6%	129
Yes	70.4%	307
Comments		93
	answered question	436
	skipped question	49

Q48. Respondent comments For Internet resourses see page 180.

1	Yananow
2	www.protonbob.com. Very helpful site for info on proton therapy and helpful with my insurance battles w/ BCBS (emphasis on their "BS").
3	Wonderful in checking medical terms, but advise reading the info on the internet with a grain of salt.
4	very little info for hifu patients
5	USTOO support group and other prostate cancer on line groups
6	Use Seedpods list - part of UsToo

7	Use Net for general support in all things where better information helps. Especially health, but many other areas. This supports me, and the Net is, obviously, a network (Sorry, I'm a bit of a techo!). I first answered yes, here, then decided that's not what you mean. I don't use it for contact with PC survivors or the Group, apart from the odd e-mail re meetings, surveys etc. (Which led me to this survey).
8	use internet for PC info., not for social networking for PC support
9	Us Too, and ProtonBob newsletters. Johns Hopkins updates
10	Us Too web site
11	US Too has been my major support group
12	US TOO GROUP PENILE IMPLANT GROUP (ON YAHOO)
13	US too and net zero are excellent helps
14	Took part in phone development program and internet as well. Cancer support group studies,
15	The US TOO!! PCAI support group is without question one of the best support groups on the internet as it relates to prostate cancer and intimacy problems. Many of the survivors who contribute to this LIST have far more practical knowledge and experience as to how to work through these issues that most specialists who work in this field . . . including people who write books about something they themselves have never experiences.
16	The Circle was/is GREAT.
17	Sometimes
18	Some
19	RESEARCH; SEARCHES; COMMUNITY CANCER OUTREACH PROGRAM
20	Research for information, emotional support.
21	read the emails and discussions in USTOO every day or two prior to surgery and about 6 months after, no longer follow.
22	Read SeedPods
23	PROSTATE VIDEOS.COM USTOO.ORG
24	prostate cancer websites
25	PCAI, Circle, USToo mailing lists.
26	P2P
27	Occasionaly for information
28	Not yet, but US TOO has just begun to seek us out.
29	Not yet
30	Not very helpful for issues relating to sex. ALL groups and information on the web ASSUME EVERYBODY is in a relationship. I might as well be a houseplant.
31	Not in need of a support group at this time...Dealing with my other cancer now
32	New letters of universities etc.
33	My close friend is the active chair or co/chair of US TOO in my region. I shared info and discussed the whole process with him .. before and after

34	Mostly I just read what others have written. I do not actively participate in any online support groups
35	MAINLY PEOPLE WITH PROBLEMS ON THE FORUMS, A BIT ONE-SIDED
36	Main source of information
37	Just started using...more of what I can contribute to others who do not have my level of experience with PC and post-op conditions.
38	jUST FOR ADDITIONAL INFORMATION
39	Joined two email forums
40	It will not make any difference. My wife does not participate nor does she care to.
41	Internet helps with communications and sharing with others who otherwise would be too far away.
42	I used the internet initially but then found that on almost every issue there were strongly opposing views and advice. I found this confusing and often depressing.
43	I use internet occasionally as an updater for information I have or to check out new procedures
44	I still monitor the Seedpods, and newly diagnosed email lists
45	I receive a lot of messages from other prostate cancer survivors
46	I read a lot at P2P, UStoo and other internet sites
47	I make use of the support group and friends I have made there a as a support network.
48	I make use of the internet as a support resource! This is mainly due to not being provided complete, truthfull information by the medical community where I live. eg: The treating physician that prescribed the anti-hormone treatment had NO interest whatsover in discussing methods to maintain sexual funciton. I also realize that there is a lot of false information there but if you have no other resource you must use what you have.
49	I maintain the Member Database and distribute email messages to the group.
50	I look for websites and research
51	I help other with their discission making when possible - I am very pro Proton Beam Treatment
52	I have tried different groups on the net - most have been discouraging, I'm pleased with pcai intimacy group.
53	I have only begun to explore the possibility of reading about ED on the internet. No really used as a support group, only exploring the possibilities that exist for a solution to my ED.
54	I have been a support person for scores of newly diagnosed men.
55	I have been a member of several on line support groups
56	I have a cousin in England, after I had my prostatectomy and mentioned to him, he went for a DRE and PSA. His PSA was 38, he also had Prostate Cancer. He received external beam radiation.
57	I get current research inf from Sebastion Amedeo This is OK

58	I find moderating a support forum online therapeutic at times, and at other times too stressful. I want to try to get off this bandwagon of prostate cancer obsession at some point.
59	I do research and keep updated by visiting some of the leading prostate cancer websites.
60	I do not need a support group. I understand my cancer and my treatment and I am at ease with both. At my age 77, I do not think I will die from prostate cancer.
61	I did research on the internet and learned that the urologist that did the radiation seed theraphy and Lupron treatment was negligent in not advising me of the effects Lupron would have on me and I believe was duty bound to do all he could to maintain my ability to have an erection.
62	I did original but after 10 years stopped.
63	I CHECK CANCER PROSTATE INFORMATION ONCE A WHILE
64	I belong to Prostrate pointers.
65	I belong to an online support group PCAI (Prostate Cancer Impotence) to help others. I have become an activist in prostate cancer issues for about 12 years now. Being able to help others has helped me inmensely.
66	I am on the US TOO website
67	I am often online looking for new information on issues I am experiencing.
68	I am a member of several support goups and read/particpate through emails
69	http://prostatecancerinfolink.ning.com
70	Heavily
71	Healingwell.com is the best I have found
72	healingwell.com
73	Have to be careful what sources you use and look for physician moderated sites.
74	Have subscribed to "USTOO" ever since, but don't always read.
75	have no idea of internet support. I would try it if I knew a link to go to.
76	Have new web site www.peiprostatesupport.com
77	go to interner for info..Us Too web site
78	For natural options
79	FOLLOWED PROSTATE POINTERS FOR ABOUT 1 YR AFTER SURGERY
80	Follow man Straight and Gay support groups. ALways looking for additianl survival information, but especially help for damaged Libido - find little discussion or help even on Lupron sites.
81	Facebook group for prostate cancer survivors
82	Email lists have been valuable.
83	Checked to see what the implications would be of bladder cancer, since my initial problem was a distended bladder due to enlarged prostate blocking urine flow significantly.

84	checked for clinical trials; follow several organizations: USTOO, PAACT, NIH, etc
85	Certainly did around treatment time. Not so much now that PSA readings are low and Drs. are not concerned.
86	But not often
87	XXX XXXX of UsT00 freguently screens and forwards pertinent info. That is how I got this survey request.
88	best source of info available, Dr. was useless
89	Belong to USTOO and use their PCAI and PC blogs
90	Almost all of my information comes from the Internet
91	Advanced Prostate Cancer support group on Yahoo. Good sounding board for issues and questions as they arise.
92	a little at first, but to detached for me
93	A fellow patient and I emailed daily during our recovery from surgery.

21% of survivors commented on their use of Internet sources.
(List of Internet cited Internet sources on page 180)

49. Have you found internet sources helpful?

Answer Options	Response Percent	Response Count
Not helpful	2.4%	10
Not very helpful	8.5%	36
Helpful	49.8%	211
Very helpful	39.4%	167
	answered question	424
	skipped question	61

Most men, 89%, found the Internet helpful or very helpful.

50. What do you see as the weakness of internet research?

Answer Options	Response Count
	292
answered question	292
skipped question	193

Q50. Respondent Comments

1	Your date entry method was not standard US and I first entered incorrect dates
2	You never really no who the people are that write or how biased they may be
3	You need to know what you are looking for.
4	You must be discerning when searching for information or you will get the wrong information.
5	You mst sort the wheat from the chaf

6	You have to understand the internet has the "fan boys" of a solution and then those who are on the wrong side of the statistics. There is no in between. As a user you must understand you are looking at extremes.
7	You have to take it for what its worth .. just take the facts.. learn from peoples experiences
8	you have to read between the lines and not take everything as gospel check everything
9	You have to be experienced enough in life (not just the Net, and not at all at being a net whizzkid!) to judge whether people know what they are talking about or not.
10	You have th=o be wary of the info as some is not trustworthy
11	You don't know who you are talking to.
12	You do not always know the background of the person providing information.
13	You can't tell what is legitimate and what isn't
14	You cannot trust all of the information on there, just as you cannot trust all of the doctors and medical staff. You must think for yourself!
15	YOU CAN OVER RESEARCH
16	work place exposure to chemicals and identified cancer sources through tissue testing.
17	widespread misinformation, unverifiable anecdotal info
18	where to find most of the material.
19	what to do
20	What one reads on a topic may not apply to his particular situation and lead to misunderstanding, unless confirmed by discussion with the doctor.
21	weeding it out
22	Web is very helpful, and can be very biased; scamers abound selling "stuff that works" for big bucks. Must reade with an educated mind, and skeptecic attention. Abstracts, incomplete reports generate hopes that are notborne out, or take much time.
23	Veracity of info and locating sites you can trust
24	VAST AMOUNT OF INFO TO SIFT THROUGH.
25	validity, too much information,
26	Validity of the source
27	validity of studies and anecdotal comments
28	validity of info-- can the source be trusted?
29	uses too many medical words i dont understand
30	unverified info
31	unvarifyable opinion
32	unreliable sources
33	Unreliable
34	Unless one knows how to drill down to the specific issue(s) at hand, internet search engines, e.g. Google, have a tendency to provide information that is too general and/or too unrelated in scope.
35	Unable to communicate readily - difficult to easily pose questions. Replies - if received at all - are vague and of little value.

36	trying to filter through all the data and determining what is truth
37	Trusting the source.
38	Trust
39	Too much to wade through. I like the monthly UsToo bulletins.
40	Too much to sift through, unreliable
41	too much spyware that causes you to get porn and such
42	too much rubbish
43	Too much information; also too general
44	Too much information. Some information inaccurate or one-sided.
45	Too much information, need to find a trusted source.
46	too much information without a moderator or guidance
47	Too much information that is technical in nature
48	Too much information and some of doubtful quality
49	Too much information and people tend to write only negative experience but not positive - so internet research usually ends-up in depression and BIG scare.
50	too much information
51	Too much information
52	TOO much info
53	Too much data, conflicting.
54	too much conflicting opinion
55	Too many varied opinions and questionable information so that is why I do not use it.
56	too many studies, not enough definitive answers
57	too many sources, which are the best?
58	Too many sources, not equally reliable or complete. Conflicting information. Some outdated information.
59	Too many people with an opinion and no science.
60	too many false positives, so to speak, that is people who pretend to have the perfect response to your problem.
61	Too many fake sites proclaiming information but just selling stuff. Also, people who had problems can be very vocal and generate excessive fear.
62	Too many ads for cancer cure etc.
63	To much miss information and false information and quackery
64	TO MUCH INFORNATION, CAN BE OVERWELMING
65	to much info and what to believe
66	To much conflicting info. To many so called expert
67	To much conflicking information
68	to many totaly different opinions
69	to detached for me
70	Time
71	There's lots of stuff out there.
72	There needs to be an agreed upon standard by which to compare information. There is too much misinformation.
73	There is too much 101-level material, and too little about the advanced stages. When Lupron stops working, what then?

74	there is a lot of information and some sources are not accurate or valid
75	Thera ar many informations without interest
76	The volume of information is overwhelming
77	The validity of the information
78	The sheer volume of available information and the time required to vet the sources for any particular bias - medical practice, etc.
79	the negativity- just give me the facts
80	The myriad of information. Get tired of looking.
81	the lack of conversation for clarity
82	The inability to ask direct questions.
83	the credibility of the source of information. a badge of approval for medical information might help
84	Teh article writers woudl seel you as great, whatever they peddled. There was almost no objective comparison of tretament methiods. I felt many of the purveyors of their own treatments to be almost unethical.
85	Takes review time to determine if web-site contents are valid
86	specifics related to own situation
87	specific case information
88	specific progresion info
89	Sorting the wheat from the chaff
90	sorting out truth and fiction
91	Sometimes superficial. Frustrating when an article is available as an abstract but not as the full article. Non-specific terms such as genito-urinary toxicity when what I wanted was specifics on impotence and incontinence separately.
92	sometimes links to unproven sources and cynical of medical profession
93	Some times too much information and questionable sources, and too scientific.
94	some things aren't as current as they could be
95	Some sites will lead you to wrong solution.
96	Some sites (particularly American) are self promoting
97	some of it doesn't have a sound basis in fact
98	some medical language problems, fear of offending someone
99	some is very anecdotal and not substantiated
100	Some information is biased by manufacturer of product or service
101	Some differences of opinions
102	sifting thru
103	Sifting through the sites
104	Seperating good information from mere opinion or uninformed comments.
105	Separating the wheat from the chaff
106	See above except from notable sources Cie) US TOO, Mayo, Etc., and to answer specific medical questions.
107	Reliable sources
108	reliable information

#	
109	Reliabilty of information.
110	reliability of info
111	Reliability
112	Reliability
113	Reliability
114	Really have not tried it much
115	Quality of information
116	Personal comments
117	People can be misled if they do not be careful about the source, the credibility, whether it is scientific or sujective/personal opinion. I have a degree in Mathematics, so I tended to most value published peer reviewed studies and papers.
118	overwhelming on small matters, lacking on the bigger questions
119	Overload. Difficult to get the large picture.
120	Opinions and information that is not up-to-date.
121	Only if don't check sources
122	One must be careful the source is reliable and accurate
123	One has to be able to differentiate good information from bad information
124	OFTEN THINGS ARE ADVERTISED THAT DO NOTHING. I MAKES ME FEEL USED
125	Often lacks context -- which my Dr and Specialist both provide, and also the support group.
126	Nothing
127	Nothing
128	not vetted by medical doctors
129	Not talking about ED
130	Not sure. One partner could fill out both forms? Both partners could be dishonest? Not all people will respond nor do all have the Internet capability, strange as that might sound.
131	not sure of validity of info... not a easy resource for dialogue
132	Not sure at this time
133	Not real life answers
134	not random
135	Not personable.
136	Not necessarily Australian
137	Not knowing the reliability of the sources of information
138	Not knowing how reliable it is
139	Not enough tied to my individual circumstances
140	not coordinated
141	Not being able to talk to the expert and now always clear explanation of what they mean
142	not being able to talk to people
143	not and opportunity for one on one feedback
144	Not always reliable
145	Not always knowing what is important
146	not always correct information, scams and "cures"
147	Not all information is correct.
148	none -

149	None
150	None
151	None
152	None
153	None
154	None
155	None
156	None
157	None
158	None
159	None
160	None
161	None
162	NO PERSON TO PERSON TALKS
163	No peer review. Uncertainty of accuracy
164	No peer review -- bullshit looks as authoritative as authentic science
165	no one to talk to directly
166	No one to talk to
167	NO MEDICAL VETTING OF INFORMATION
168	no follow up
169	No focus; validity concerns
170	No central clearing house for treatments or alternative treatments
171	NEVER KNOW THE RELIABILITY OF INFORMATION PROVIDED
172	Needs to be evaluated critically (if that is a weakness)
173	Needs more clinical application
174	My type of problem is not addressed. If it is, not easy to find.
175	my lack of knowledge to navigate systems. It's still rather new to me.
176	My lack of interest. I've accepted my present condition as "what is, is "
177	Must weigh the reliability of the information. Is it a reputable source? What's their agenda?
178	Must be very careful in trusting
179	much is anecdotal
180	much conflicting information I rely more on books written by doctors,phd's. and survivers
181	Much bogus info out there by people trying to make money
182	Much anecodotal information not supported by fact
183	Most support groups as with urelogists do not mention Proton and it's benefits
184	Most information found on the internet has been peer reviewed by medical professionals . . . but that is not always bad!!
185	Mone
186	Mis-information.
187	Maybe I do not know how to use it. Found a website that has Partin Tables and some other info that seams interesting to me but they ask for $49.95 to send the info. I am struglig saving money to pay $636.00 on Property Taxes. So that info is going to wait.
188	may be opinions and not factual

189	Material on the internet is not filtered. Need to be careful making sure that web sites have reliable and accurate information -
190	there is a tremendous amount of emotion conveyed about prostate cancer. For some people, I can see that it may be difficult to evaluate objectivity of the information being conveyed via the internet
191	lots off bad information
192	Lots of stuff but not easily located
193	lots of information in many places, need to condense
194	lots of conflicting information
195	Lot of unreliable, junk information to sift thru
196	Lot of sales for various itmes. Rip off.
197	Lot of junk science and fraudulent claims that are difficult to identify
198	Legitimacy of websites
199	Lack of specificity
200	lack of info
201	lack of control
202	Lack of ability to raise questions and get feedback
203	Knowing where to start.
204	Knowing where to find it. There needs to be a better way to "advertise" its existence.
205	Knowing what to believe
206	knowing what is true and what is not
207	it requires ability to interpret clinical data, which many people don't know how to do
208	It is often non personal and acute, patient specific information is hard to find
209	It is hard to find the right answer for your specific questions
210	It is far away. Rarely speaks to my immediate situation.
211	it does not hlep you decide much whether to proceed with radical prostectomy
212	It can be difficult to seperate truth from fiction.
213	It assumes that men without partners don't exist
214	Isn't specific to my questions
215	is the information reliable and based on research
216	Is the information factual
217	Internet search will not assist me with my wife
218	Interaction
219	Insuring the veracity of information.
220	Information quality variable
221	Information overload
222	Information overload
223	Information not always correct
224	INFORMATION IS NOT SCREENED FOR ACCURACY
225	Information can be overwhelming and not always dependable
226	Ineffective, lacking specific data.
227	Inconsistent quality.
228	Inability to see biases and accuracy of information
229	Inability to check source of funded research. Must be careful to ensure research is not biased toward certain treatments.

230	Important to ensure the website has reliable information.
231	Impersonal, not as interactive as in-person contact.
232	Impersonal for the most part
233	Immediate availability of information
234	If I come across some informative data, I tend to forget to bookmark the site. Also, there is just too much information available and it confuses me as to really where to look for help.
235	I think it's a big help, great way to get answers
236	i need to interact with other med with pca
237	I have no results
238	I find the internet too commercial.
239	I don't have the patience to go from link to link or to read either on line or printed
240	I can't really compare it to other forms of support as I haven't spent much time in one on one groups.
241	I heard the same complaints about erection problems, leakage and other problems that I didn't have ; so I just didn't stay . I go back occasionally, but I don't conttact any other survivors.
242	how does the inexperienced PC man filter through all the material available on the internet. Lots of crap.
243	Having to pay for some of the information, I do not believe that a medical facility should charge for info.
244	Have no opinon
245	Harder to find current info for "aggressive prostate cancer"
246	hard to tell how accurate they are
247	Hard to find specific info
248	hard to find sometimes when talking about sex leads to inapporiate websties
249	Great deal of information to sift through and determine what is approptiate to my situation
250	First, there is only limited full text access to journals. Secondly, there are too many quacks and their followers out there.
251	Finding information that you want
252	Finding authoritative, experienced people in the same situation
253	Finding a situation similar to mine
254	figuring out what fits me
255	factual data lack of credtiability
256	everyone is different with thier cancer
257	Everyone has a diffrent idea, and they don't like the others!
258	Effort needed to locate "complete" info...you see only info you are able to find...
259	don't know how reliable source is
260	Doesn't address issue related to this survey.
261	does not addrees inconsinence
262	Do not always get the correct info or am not asking the right questions
263	difficulty in separating the 'science', 'politics' and 'advertising' inherent in the information provided

264	Difficult to find answers to specific questions, information is general in nature.
265	Dialogue too lengthy and clinical for a non-medically trained person
266	Depends on the website. Mostly always helpful
267	Depending on the site, one can get spammed.
268	Danger of inaccuracy
269	Credibility
270	Credability
271	confusing informations
272	conflicts in treatment
273	conflicting views
274	Conflicting information.
275	Conflicting information from site to site; lack of active chat-sessions with qualified medical experts
276	Confidentiality
277	Commericial intrerests that color information and opioions
278	CANNOT ALWAYS FIND ANSWER TO QUESTIONS
279	Can never be sure of the accuracy of information/
280	Breaking it down into understandable language, conflicting information
281	Biases are not always clear. Currency varies, but is not always clear. Lack of comparative info
282	Beleiving everything you read
283	At time of surgery info on cryosurgery was lacking.
284	artiles not clearly dated (and some are outdated)
285	articles for medical profession/ some sites selling services or devices
286	An overwhelming number of sites to wade thru many with th same perspective.
287	almost too much; newly diag. man has prob. discerning the truth.
288	accuracy, contradicting information
289	access to medical journals not permitted
290	Access to medical infromation that I want is sometimes restricted by membership requirements/cost
291	?
292	(and you and I are doing this over the Internet?) Bogus and mis-information from comercially sponsored sites.

65% of survivors listed the limitations of the Internet as an information source.

The survivor respondents are highly educated, which would tend to bias the answers.

51. What do you see as the strength of internet research?

Answer Options	Response Count
	296
answered question	296
skipped question	189

Q51. Respondent Comments

1	YOU HAVE TIME TO RE-READ
2	You hae to be careful to find reliable information and recognize anecdotal information for what it is.
3	You find out what your doctor isn't telling you.
4	You can quickly investigate any subject.
5	You can look for answers to questions you don't want to voice.
6	You can find scholarly articles or sites put out by cancer centres
7	you can do it in your own time
8	you can check on your own and on your own time.
9	Worldwide nature
10	wide variety of established researchers
11	wide range of opinions, vast resources, up to date, impersonal
12	Wide range of access to information. For example, after reading Patrick Walsh's book I found much usefule info on the Johns Hopkins web site.
13	Wide band of institutions and lack of worry about intimate details.
14	Wealth of information through subject searching.
15	Wealth of information - many testimonials from survivors
16	wealth of info
17	Volume of information
18	volume of information
19	volume of information
20	Very informational
21	Very available on my schedule with many sources.
22	Vastness
23	Vast quanties of information
24	Vast body of Knowledge
25	Vast amount of ingo available on prostate surgery, recovery etc.
26	variety of potential sources of valid information
27	variety of information relative to my problem.
28	Up-to-date information on the disease; some site provide information at a patient-level (non-technical) yet effective; easy to collect information from survivors that can be beneficial to others
29	Up to date info
30	unvarnished open social networks with real stories
31	Unlimited availability

32	Unfettered comments of survivors
33	Timeliness, scope and array of information available.
34	This format is excellent. Hopefully the outcome this study will be a properly administed Q and A Internet Site.
35	there when you need it
36	there is hope
37	There is good information if carefully assessed
38	There is all the information you want if you are willing to spend the time.
39	There is a wealth of information about one's options when one is first diagnosed.
40	There is a LOT of information and a Lot of misinformation
41	There are many helpful sites and things can be shared by fellow PCa patients.
42	There are alot of various opinions and suggestions
43	The volume of information is overwhelming and is so available
44	The volume of information available.
45	The timely scientific data from research and studies.
46	The number of sources and people using them
47	The is a lot of information out there
48	The information is abundant and mostly clear.
49	The broad reach and the ability to connect with people who have expertise.
50	The amount of information out there
51	THE AMOUNT OF INFORMATION MADE AVAILABLE
52	The amount of information available - although it too can get discouraging to find the right site, and to deal with sites designed to advertise. I need a "map"
53	the amount of information available
54	the accessibility to a larger population
55	the abundance of resources
56	The ability to hear from others who have actually gone through various treatments
57	The ability to follow discussions with many people with all forms of treatment.
58	Supposed to be free and at the reach of everybody. I just have to learn how to use it.
59	Support
60	Stays very current (up to date)
61	Speed of searches. Availability of more than I would have found had the internet not been available and I had to talk my way into a medical school library.
62	Speed
63	Specifics
64	Specific and in depth info.
65	source of new treatments
66	source of comfort and ability to educate yourself

67	something you can do at home
68	Some web sites such as UsTOO and American Cancer Society are helpful because they are objective.
69	Some sites provide excellent, factual information.
70	Some info on the Internet is bogus
71	sites staying current on info
72	shows most recent data first
73	should be current info and you do not need to leave your home to get such
74	self awareness, private and at the persons pleasure
75	Seedpods is an internet support group more than research although it allows you to ask questions men who have has similar experiences.
76	See above. Since you are dealing with extremes you can not only better your options by seeing how those that have failed are being delt with by the medical community and we who are doing fine have a reference point for how well we are doing.
77	SEARCH FROM HOME; CORRELATE STATISTICS AND DATA; BOOKMARKS; FAVORITES; EMAIL
78	responsive to individual needs
79	Research any time from anywhere.
80	Reports on the latest research
81	Real person accounts
82	READ OTHERS' EXPERIENCES
83	Range
84	Quick and easy
85	Quick access to vast amount of information
86	provides lots of background info to use for doctor questions
87	provides a wide variety of sources that are easy to access
88	privacy....good education..
89	Power. Huge information resources at your fingertips, and fast - compared with pre-Net days - even after allowing for time spent sorting through dross.
90	plenty of information on options
91	Plent of it
92	Personal stories and personal advice from so many that had been thru the successes and failures of their treatment choices.
93	people tend to be open and candid
94	Outstanding access to current and valuable studies and research
95	Options
96	One can research it at your lesure and throughly once you get into it.
97	Number of testimonials
98	Not sure at this time
99	Non-typical questions were answered by others who had similar symptoms (bleeding 5 weeks after surgery)
100	None
101	new procedures and info in general
102	New findings
103	much personal opinion

104	Much information available if you know what to look for
105	MUCH INFO QUICKLY
106	Much better information than I ever got from any Urologist and I have had many searching for a good one
107	Most topics are addressed
108	More is better.
109	More honest structure.
110	men with pc can connect and discuss problems, issues and whatever is bother them.
111	Medical information and describing treatments.
112	med info, mostly info from fellow victims
113	Massive amounts of knowledge
114	Many sources, very available, possible to cross-reference
115	Many sources of information of prostate treatments
116	Many medical based sites have solid relevant information and links to other good sites
117	Many knowledgeable peole and docs
118	many different resources
119	many different point of view
120	many available sources,sometimes to many
121	Lots to read and see. May be very up to date.
122	Lots of options available
123	lots of information for someone, like me ho wanted as much info as possible
124	Lots of information at my fingertips--University/hospital web sites offer a wealth of information
125	Lots of information and different views.
126	lots of information
127	lots of information
128	lots of information
129	lots of information
130	LOTS OF INFOR
131	lots of info. out there
132	lots of good info
133	lots of free information
134	lots of different information and views to review
135	lots of anonymous information
136	Lot of information
137	Links to current research
138	limited only by knowledge & ability
139	Learn about some issues not discussed in support groups or health care providers
140	latest research available immediately
141	Latest encouraging treatments especially natural helps. Also details to help in physical problems.
142	Larger sample population
143	Knowing what to believe.
144	knowing others are out there
145	Keeping up to date and learning from others

146	It's easy and you can study alone and at your own pace
147	It's available 24/7. If you have patience you can find information on almost everything related to prostate cancer.
148	its available 24 hours a day
149	It should provide suggestions and help for all symptoms and problems expressed by prostate cancer victims
150	It should be able to reach a very large population.
151	It makes information more readily available.
152	It is better than nothing. And can give you literature references to do your own literature search.
153	It is available
154	It can reach a lot of people
155	It allows you to anonymously research just about anything you could imagine
156	IT ALLOWED ME TO COMPLETE THS SURVEY-THANKS
157	Internet research provides information that is less political or biased
158	International, up to date
159	Informative
160	information of 1º Class Urologists and Oncologists (Labrie, Strum, and many others)
161	Information is far more readily accessible and faster to retrieve than trying to ferret out specifics from hard copy sources.
162	Information at your finger tips
163	Information
164	Information
165	Info can be found from home
166	Impersonal
167	immediately available information
168	immediately available
169	If published by a reputabal enity like the Mayo Clinic, it's easier to "digest."
170	If one pick the appropriate source, the info can be very valuble.
171	I was able to go to the Surgeon offfice much more informes about what I was about to undertake.
172	I could see that I was not alone, especially in my urinary problem at "climax".
173	I could find something that my doctor is not even aware of.
174	I can really dig deeply into questions
175	I can find specific topics.
176	Huge amount of information easily available. Much of it excellent and difficult to access w/o internet.
177	How everyones journey can differ
178	Hospital portals
179	helped me decide which treatment to have
180	help to answer questions
181	Have no opinion
182	has lots on information for free
183	Groups like yananow.net provide access to large numbers of men
184	great deal of information including survivors stories

185	Good quality information from groups like ASCO (Am. Society of Clinical Oncology), JAMA, etc. Access to databases like SEER.
186	Giving out support information
187	Gives tremendous access to information that otherwise would be very difficult for the lay person to come in contact with. It provided me with far more information than I could ever have obtain from any of my doctors.
188	getting multiple second opinions, acessibility to many experts across the country
189	general information
190	Gain information you would not otherwise have
191	finding the right doctor/surgeon
192	finding out what is new
193	Find out new treatments, medicines, trials very quickly.
194	Find large volume of information quickly
195	fast information
196	fast and you don't have to leave home
197	extent of information available
198	Exposure to various types of treatment, latest research, as well as different points of view.
199	Excellent source of information (Sturm); ability to challenge urologist who insisted on surgery
200	Excellent information on all treatment options in a neutral way. I found that the various urologists were all biast to one or the other treatment options. Social networks like the new prostate cancer info link are extremely helpful in getting information and getting 'positive' views in the internet.
201	e-mail information
202	Educational and informative
203	easy to use from home
204	Easy to use and confidential
205	Easy to search various subjects
206	Easy to find lots of data and information
207	Easy to find information.
208	Easy to access
209	Easy of obtaining information
210	easy and private
211	easy and fast access to a world of knowledge
212	easy access to varied information, connectivity with others with the same disease
213	Easy access to information
214	Easy access
215	easy access
216	Easy
217	Easy
218	Easily find people who have same problem as me
219	easily available, lots of information, some very cutting edge info, videos, discussions
220	ease of use

221	ease of searching and availability of up-to-date inforamtion
222	Ease of obtaining information
223	ease of obtaining data
224	ease of access
225	ease of access
226	Ease of access
227	Ease
228	DETAILS FROM DIAGNOSIS TO POST OPERATION,ETC
229	Depth of info available, provided it is readily locatable...must be well cross referenced to assist search effort...
230	cutting edge information, broad pool of source material
231	Current information, better able to assimilate the information in a less traumatic environment ie: doctor's office
232	current information, reliable information from some sites ie Mayo clinic, Canadian Cancer Society
233	Current information, rapid access, hope for PC victims
234	convienent, quick
235	convenient accessability
236	Convenience of hearing first hand experiences
237	connected me with other men and their partners - who are open, frank and fearless
238	comprehensive if you are willing to sit there and reseach
239	Choices and confirmation,
240	Checking medical terms and at times the various treatment results. Also locating support groups
241	Can find information on variety of subjects dealing with prostate cancer.
242	can find good organizations and follow scientific meetings;
243	can find answers you probably wouldn't ask or forgot to ask your doctor.
244	can find all kinds of information for issues.
245	CAN ASK AS MUCH AS YOU WANT
246	broader topics in your home readily available
247	broad resource
248	Broad range of experiences
249	Broad diversity of subjects
250	broad base of available information
251	Breadth of information
252	Being up-to-date on what some reliable sources have researched, done.
253	Being able to communicate freely, able to validate my own thoughts and feelings, comradship, invaluable.
254	Being able to access results of recent scientific articles about prostate cancer, keeping track of emerging treatments, getting a broad perspective on attitudes and policies of patients, caregivers, prostate cancer non-governmental organizations, and the medical community.
255	Availabliity
256	available when you need it
257	Available info good as long as you are on the right site.

258	available in lieu of face to face support
259	Available facts
260	Available anywhere anytime. Some inhibitions which may be present in person disappear on-line.
261	Availability, amount of information.
262	Availability of virtually unlimited amouts of information.
263	Availability of info on treatment alternatives.
264	availability of info
265	Availability and a broad range of points of view. You are not relying on one doctors opinion or personal preferences.
266	Availability 24X7.
267	Availability
268	as above
269	answer to questions or concerns that come up between doctor or group meetings
270	Anonymity
271	and awful lot of information and not all in agreement with each other
272	An incredible amount of information.
273	Always accessable, varied views, personal and medical information, interactive.
274	alternative treatment possibilities are presented that most doctors are not aware of
275	alternative info
276	Almost the same as above, e.g., lots of information.
277	almost anything and everything to check out, but must have help sifting it out.
278	All the information you want to know and need to know it at your touch
279	All possible treatments are described with historic data on success rates etc
280	accessibility to anyone with a computer and a librarian's instincts for research
281	accessability
282	Access top research and experts
283	Access to ortherwise unavailable info
284	Access to online groups of actual survivors.
285	Access
286	abundance of data
287	Able to talk and offer or recieve help wth questions etc.
288	Ability to search for data defined by the Google search terms.
289	Ability to make informed decisions based on scientific research results
290	Ability to get a lot of information from a variety of sources
291	Ability to access information from experts
292	A plethora of data
293	A lot of information available that you don't have time to discuss with the doctor.
294	A great deal of medical info. I was much more informed by such before going to my physician

| 295 | a flood of information |
| 296 | ???? |

66% of the survivor respondents commented on the benefits of the Internet. The survivor respondents are highly educated, which would tend to bias the answers. The comments included accessibility 24/7, speed, up-to-date information, depth of available information and wealth of information.

52. Do you have an informal support network?

Answer Options	Response Percent	Response Count
Wife	71.8%	254
Friend	39.8%	141
Prostate cancer survivor	53.7%	190
Other	15.3%	54
Other (please specify)		87
answered question		354
skipped question		131

Informal support networks are important. Partners and families top the list. Other survivors also provide important support.

Q52. Respondent Comments to "Other (please specify)"

1	A leaderless Mens Group for past 15 years.
2	adult children
3	big family support
4	Brother
5	CHAPTER LEADER, PCa SUPPORT GROUP
6	Children
7	Children
8	Children
9	Church
10	Church ... Rotary ... long friendships
11	Cousin
12	Extremely helpful and engaged urologist
13	Family
14	Family
15	Family
16	Family
17	FAMILY DOCTOR
18	friends from within my support group
19	Have friends that comiserate together with same problems
20	have had some discussions with three to four others
21	healingwell.com
22	Cancer survivors Monthly meetings
23	I provide support for others

24	I'm very analytical and very much self contained. I'm at the end of my 7th year as a survivor and prostate cancer, while still very real for me, has become much less so for family and friends who believe I'm cured. If I talk too much about it, they tend to withdraw. I believe this is a natural reacion.
25	In the begining wife and family were there, but have fallen away - I think cancer is a very isolating disease eventually for most
26	Internet chat group
27	Just the internet exchanges
28	Kids
29	many PCa friends thorough the years
30	massage therapist..male
31	Monthly support group, two friend/colleagues at work, one of my brothers
32	My adult daughter and son are well informed and supportive
33	my male partner of eleven yrs
34	My male partner. I'm gay, remember?
35	My men's group (not specific to illness)
36	my partner of 16 yrs
37	My wife and grown children are my support
38	My wife died, I have no friends or acquaintances so no network
39	My wife has been my only real support.
40	Myself
41	no
42	NO
43	no
44	No
45	No
46	no
47	no
48	no
49	No informal support nework
50	no support group.
51	none
52	None
53	None
54	None
55	None. My wife was really never supportive. She has now left me as of 2 months ago. Although we still speak on a daily basis. But the conversations are just small talk, e.g., how's the weather (she's out of state in WI, I'm in AZ), what did you do today, etc. etc. When confronted as to why she was leaving after a long and, what I thought, was a great relaionship, she stated: "You got cancer, then got depressed. Because of your depression, you got me depressed and I don't want to live with you anymore." - that's almost verbatim.
56	not other than the group
57	other cancer survivor

58	other men I went through treatment with
59	Our local support group
60	Our Voice
61	partner
62	partner and family
63	PCa online support groups
64	physician
65	Prostate Cancer Association
66	Prostate Cancer Support group through work
67	Prostate Survival Group
68	PSA Support group when I visit them now I have been 3 years PSA .001
69	Psychiatrist
70	Psychologist
71	PCA SUPPORT GROUP
72	Regular sessions with a clinical psychologist to help with depression.
73	Relatives supply moral & emotional support. Physical help impracticable, due to distance
74	Some of my children and grandchildren
75	SUPORT GROUP THAT I CHAIR
76	support group, learning to take control of own health care
77	surviving brother that had prostate surgery
78	Traditional community
79	US Too local chapter
80	US TOO Support Group
81	UsTOO survivor and ed. group
82	Very close relitive works in Uro office and is a PC Serviver
83	volley players; large family
84	We organize a motorcycle ride to raise funds for prostae cancer research, education & support
85	When I try to bring these issues up, people are unresponsive or offended.
86	wife is a nurse
87	YANA, New Prostate Cancer Network

53. How beneficial has the informal support group been?

Answer Options	Response Percent	Response Count
A hindrance	0.3%	1
Not helpful	3.1%	11
Somewhat helpful	21.8%	77
Slightly helpful	15.3%	54
Very Helpful	59.5%	210
If you checked a hindrance please explain		8
	answered question	353
	skipped question	132

53. Respondent Comments to "If you checked a hindrance please explain"

1	After the initial scare she has seen that I am still feeling fine with absolutely no problem from the cancer or from the proton beam treatment. She now thinks I will be around for a while.
2	couldn't have kept a clear mind without him
3	I don't see how anyone can survive PCa without interacting with others with the same condition.
4	I don't think I could have made it through the depression without her support.
5	It has made me realize that it it up to me to investigate more thoroughly and to ascertain a MD's qualifications and treatment recommendations.
6	Same as in #51.
7	The encouragement to "do something" and not just sit on your hands and complain.
8	wife is sympathic and supportive but I don't think she realizes how important it is for me to be able to perform sexually.

Informal support is very helpful for 59.5% of the and only 3.1% found it not helpful.

54. How do you prefer to get prostate information?

Answer Options	Never	Seldom	Occasionally	Often	Always	Response Count
Email	57	28	108	102	53	348
Blog	180	36	36	14	9	275
CD	176	41	31	5	5	258
DVD	148	42	44	14	6	254
Web Video	153	41	44	14	8	260
Magazines	87	62	99	38	16	302
Newspapers	80	86	90	19	7	282
Library	115	66	56	22	12	271
Internet	17	17	82	178	103	397
Support Group	74	43	64	87	50	318
Facebook	223	11	12	5	4	255
Twitter	241	8	3	0	3	255
Doctor	10	34	138	112	85	379
Prostate Cancer	40	43	120	86	47	336
TV	125	93	47	5	7	277
					answered question	438
					skipped question	47

23.5% preferred getting their information from the Internet; 19.4% from their doctor, 12.1% from their email and 11.4% from support groups.

55. How do you rate your confidence that you can get and keep an erection? WITH NO MEDICAL AIDS OR ASSISTANCE!

Answer Options	Response Percent	Response Count
Very Low	61.5%	256
Low	17.8%	74
Moderate High	13.0%	54
Very High	7.7%	32
	answered question	416
	skipped question	69

56. When you had erections with sexual stimulation, how often were your erections hard enough for penetration? WITH NO MEDICAL AIDS OR ASSISTANCE!

Answer Options	Response Percent	Response Count
No sexual activity	19.6%	80
Almost never or never	35.5%	145
A few times (much less than half the time)	7.3%	30
Sometimes (about half the time)	7.6%	31
Most times (much more than half the time)	11.2%	46
Almost always or always	18.8%	77
answered question		409
skipped question		76

This data needs to be analyzed by treatment modality and by age.

57. During sexual intercourse, how often were you able to maintain your erection after you had penetrated (entered) your partner? WITH NO MEDICAL AIDS OR ASSISTANCE!

Answer Options	Response Percent	Response Count
Did not attempt intercourse	30.4%	125
Almost never or never	26.3%	108
A few times (much less than half the time)	8.8%	36
Sometimes (about half the time)	7.3%	30
Most times (much more than half the time)	9.2%	38
Almost always or always	18.0%	74
answered question		411
skipped question		74

This data needs to be analyzed by treatment modality and by age.

58. During sexual intercourse, how difficult was it to maintain your erection to completion of intercourse? WITH NO MEDICAL AIDS OR ASSISTANCE!

Answer Options	Response Percent	Response Count
Did not attempt intercourse	33.3%	136
Extremely difficult	24.0%	98
Very difficult	6.4%	26
Difficult	5.6%	23
Slightly difficult	11.2%	46
Not difficult	19.6%	80
answered question		409
skipped question		76

This data needs to be analyzed by treatment modality and by age.

59. When sexual intercourse was attempted, how often was it satisfactory? WITH NO MEDICAL AIDS OR ASSISTANCE!

Answer Options	Did not attempt intercourse	Almost never or never	A few times (much less than half the time)	Sometimes (about half the time)	Most times (much more than half the time)	Almost always or always	Response Count
For you?	124	91	41	27	46	82	411
For your partner?	96	88	40	31	52	55	362
answered question							411
skipped question							74

This data needs to be analyzed by treatment modality and by age. The survivors' "guesstimates" of their partners' satisfaction need to be compared to the partners' replies. Survivors and partners need to be matched and the replies and guesstimates compared.

60. How do you rate your confidence that you can get and keep an erection? WITH MEDICAL AIDS OR ASSISTANCE!

Answer Options	Response Percent	Response Count
Very Low	33.5%	121
Low	16.1%	58
Moderate High	24.4%	88
Very High	26.0%	94
answered question		361
skipped question		124

This data needs to be analyzed by treatment modality and by age.

61. When you had erections with sexual stimulation, how often were your erections hard enough for penetration? WITH MEDICAL AIDS OR ASSISTANCE!

Answer Options	Response Percent	Response Count
No sexual activity	18.5%	66
Almost never or never	22.2%	79
A few times (much less than half the time)	8.1%	29
Sometimes (about half the time)	7.3%	26
Most times (much more than half the time)	15.7%	56
Almost always or always	28.1%	100
answered question		356
skipped question		129

This data needs to be analyzed by treatment modality and by age.

62. During sexual intercourse, how often were you able to maintain your erection after you had penetrated (entered) your partner? WITH MEDICAL AIDS OR ASSISTANCE!

Answer Options	Response Percent	Response Count
Did not attempt intercourse	26.3%	93
Almost never or never	18.9%	67
A few times (much less than half the time)	6.2%	22
Sometimes (about half the time)	7.1%	25
Most times (much more than half the time)	15.3%	54
Almost always or always	26.3%	93
answered question		354
skipped question		131

This data needs to be analyzed by treatment modality and by age.

63. During sexual intercourse, how difficult was it to maintain your erection to completion of intercourse? WITH MEDICAL AIDS OR ASSISTANCE!

Answer Options	Response Percent	Response Count
Did not attempt intercourse	27.4%	96
Extremely difficult	19.4%	68
Very difficult	3.7%	13
Difficult	7.1%	25
Slightly difficult	14.8%	52
Not difficult	27.6%	97
answered question		351
skipped question		134

This data needs to be analyzed by treatment modality and by age.

64. When sexual intercourse was attempted, how often was it satisfactory? WITH MEDICAL AIDS OR ASSISTANCE!

Answer Options	Did not attempt intercourse	Almost never or never	A few times (much less than half the time)	Sometimes (about half the time)	Most times (much more than half the time)	Almost always or always	Response Count
For you?	95	63	32	32	61	74	357
For your partner?	77	61	34	21	64	62	319
answered question							357
skipped question							128

This data needs to be analyzed by treatment modality and by age.

65. General comments/feedback about this survey

Answer Options	Response Count
	149
answered question	149
skipped question	336

Q65. Respondent Comments

1	You asked no questions about the emotional "value" or "quality" of the relationship, either before or after treatment. Mine was/is "low," and that has had an important effect on my sexuality before and after treatment.
2	1. SOME QUESTONS DID NOT PROVIDE A COMPLETE RANGE OF ANSWER OPTIONS (I.E., YES/NO) 2. WHEN AND HOW WILL THE SURVEY RESULTS BE MADE AVAILABLE? THROUGH EMAIL TO SURVEY PARTICIPANTS?
3	1. Better than half the men in my support group - even those over 65 - tell me they do not have much of an erection without taking the sidenfal class medications. Those who do take these type of drugs generally report good results; of course, that's why they take them. Yet very few of us - and I have the resources to pay for these meds out of pocket - can afford the cost. Thus, we go without sex more than we, or our spouses, would prefer. Under my current insurance - BlueShield - I get 6 Cialis per month with a prior authorization. The PA reduces the cost from $115.00 for 6 20mg. pills to $45.00 for 6/20mg. The sidenfal like meds. have an even greater importance than facilitating individual occasions of sexual penetration. More importantly, the medication offers nerved damaged survivors to have normal morning and even spontaneous erections. This is certainly true for me, and many other men tell me they have similar experiences. This ultimately more important than individual sexual interactions! Why? Because having a normal erection has a tremendous effect on a man's sense of sexual identity and potency. It imparts a sense of psychological wholeness which, in the absence of the medication - ergo an erection - one feels otherwise impotent. Fortunately, I am married and get support and encouragement from my wife. However, lacking a secure and supportive relationship with a mate, I can envision few men who would feel comfortable initiating a potentially intimate relationship based on an occasional pill. In short, prostate cancer survivors who are fortunate enough to have sufficient nerve function to support an erection using sidenfal like medications, should (must) have access to the drug on a daily basis for it to be truly effective and for the patient to receive the greatest benefit. 2. I spoke with several doctors before opting for a form of treatment. Several Urologists addressed the fact that I would lose most erectile function following surgery, and that rarely if ever did one have escape some permanent peripheral al nerve damage and corresponding dysfunction. However, none of the physicians I consulted a single doctor ever discussed shrinkage or diminished size that accompanies the surgery. This came as total shock and to me and still bothers me a lot.c
4	A complicating factor is that subsequent to cancer treatment I also developed Peyronies Disease.
5	after 6 months, still unable to get full erection even with medication (i have tried all 3 pills)

6	After having a highly active sex life it has been an interesting time not being able to have a natural erection most of the time and then, when I do, it is not that "full" and difficult to maintain. I personally feel my libido has declined as well. I think it is critical to have an understanding partner who will experiment and will try different things to make both of us satisfied with the outcome.
7	After my surgery I tried Viagra but the side effects were distracting and didn't produce much of an erection. I used a vacuum pump that produced a erection that permitted intercourse and used that for years. About 2 years ago my wife had her own health issues and we stopped sexual activity. Now the pump doesn't work for me, but my wife is not interested in sexual intercourse (never really has been). About weekly or every other week we engage in mutual masturbation to climax for both of us. I doubt that I could get an erection at all. We are both 69 and things are great beyond the bedroom. For our entire married life, my wife would not engage in conversation with me about our sex lives. She would not complete this survey.
8	After the first three months I was able to maintain an erection for two to five minutes with a lot of help from my wife the time was expanded up to 20 to 20 minutes. At this time I was using lavetra. As time has progressed I no longer have needed to use the medication.
9	After the first year (post-operative) I started to get erections and enhanced them using Viagra. The differences between sex prior to prostatectomy and post were a huge shock ! The surgery had reduced the size of my penis - erections were not si strong - orgasm was "strange" and sometiomes painful (I could feel nerve pain) and the lack of the ejaculatory sensation was disappointing. Most dramatic is my loss of libido - I think largely because I remain incontinent the whole issue of sex - desire and comfort - is diminished. We had a great sexual relationship - vigorous, creative, frequent and important to both - and I know that I have subliminated my sex drive due to three main issues: a) incontinence b) penile size and strength c) loss of orgasmic sensation (meaning the sensation I experienced prior to surgery -)
10	Again, I think it is important that the doctor visit with both husband and wife. I think my wife may have been more interested if she realized that my sexual desires were normal for a man of my age (65) at the time of surgery.
11	Although, I have know most of my adult life that I had strong homosexual tendencies I was able to have successful intercourse until a few years before my diagnosis when my wife essentially lost all interest in sex. I was forced to resort to masturbation only to satisfy my sexual need but I seldom had any problem achieving orgasm.
12	anxious to find a way....... hope you succeed in helping us.
13	As a gay man after surgury, I have not tried penetration or intercourse. I do not use devices nor medication nor stimulation, per se. But I do have solid, long lasting erections and (dry) orgasams

14	As stated in the survey I had a duel implant in XXXX, that same year, after almost 30 years together, my wife left me. I think I was partly at fault as I had sunk into a deep depression over the fact that I was about 95% incontinent and 100% impotent, 3 years after surgery. At the same time my PSA started rising and I under went 36 radiation treatments. Recently my PSA started to go up again and I just finished my first round of Lupron. I have had one sex partner since and we were both satisfied with the implant.
15	At the beginning of treatment intercourse could be maintained with medications. Then the meds stopped working. We have had little luck with a pump. Therefore, foreplay and mutual masturbation are the result in sex. Now, ten years later, I have little interest as does my wife. Her medical issues compounded with mine has pretty much taken the desire away. The hormone suppressing meds, Lupron, has been the major issue for this
16	At the time I was diagnosed with PCa I had a partner and she was somewhat helpful during that initial diagnosis. She was extremely disturbed and upset about the diagnosis as she had an uncle that died 6? months after diagnosis. I also had and still have low libido, possibly due to a benign tumor on my pituitary. Currently I have no partner and very little interest in sexual activity and I am fine with that. I have a number of female friends including my ex-partner.
17	At the time when I found out that I might have prostate cancer, my creative energy slumped to the point where I could not follow through with any of my visual art projects although I was still able to maintain my writing. My main concern was to go through with the surgery for my family's sake and of course my own not realizing the psychological impact of not being a sexually functional male with a very short penis. After my operation I tried to get back into my art with hardship until I hit by chance on the notion to draw my own penis very large with its malfunctions in ones face as it were. These drawings I have shown other people and in fact have shown one in a gallery. I was then face to face with my demons. I don't know if this could be a help to anyone else as I would think that there are a lot of men out there who are suffering quietly with their loss. From my own experience, I would have a tendency to let other men know that what they need to do is expose themselves to themselves and others as well through whatever means they have to work with. It will seem hard at first but once they let the cat out of the bag, it will become easier and they will feel the burden lessoning. Thank you for your survey.
18	Attempts at sexual intercourse also complicated by wife's back problems.
19	Because of my sexual orientation it was difficult to answer some questions

20	*CAVERJECT WORKED UP TO THE TIME THAT THE PENIS DEVELOPED A BEND TO THE LEFT MAKING IT DIFFICULT IF NOT IMPOSSIBLE FOR PENETRATION. Am currently participating in vacuum therapy to try and rehabilitate the penis and have been advised to no longer use caverject, followup appointment will be scheduled in about three months.*
21	*Comprehensive*
22	*Cryosurgery gave new meaning to the term "knock your dick in the dirt" cause it is deader then a doorknob. Only a shot gives me an erection and it's too uncomfortable to have sex with it. A pump is meaningless so basically we just go with out. appreciate the survey though as it does give me an opportunity to address this with the wife*
23	*Despite the fact that one of the first question was wether I was striaght or gay, the survey question were very heterocentric.*
24	*do not use medical aids as of yet.............maybe to early for me to know all this for now...that might be down the road for me.*
25	*During the preliminaries preceding intercourse, there is an uncontroled leakage of urine from my penis. Although th amount is small, that is a turn-off for my partner, especially compounded by other "distractions" when we interrupt the preliminaries to use of the pump in order to get a more sustained erection. I have yet to find something to read on this subject and what can be done about it.*

26	Excellent survey; I look forward to the results. In my situation, the problem is not getting an erection, but sustaining it until my orgasim. The penal ring solves this problem, but I know that I have to have an orgasim before thirty minutes, so it limits us somewhat. I have had 6 relationships of varying lengths since my treatment, and both my partner and I have been able to have a very active and enjoyable sexual aspect. However, first and foremost, I always choose a relationship with people who also value intimacy, as I do, and the sexual apsect is just a fantastic bonus. I believe a big part of this has been my enjoyment of giving oral sex, witch all of these partners have enoyed tremendously, so I do not have doubts about whether they are also "feeling fulfilled/satisfied". I have come to conclude that intimacy, foreplay and oral sex are valued far more than penile vaginal penetration (or maybe I have just been lucky enough to have had relationships with mature, enlightened woman of like mind). Lastly, if it comes to the point where the oral medicines no longer work for me, it is nice to know that I have the trimix injections standing in the wing, and when no aids work, i know that with the right partner,we should still always be able to have a very active sexual aspect to our relationship and enjoy orgasims together. If you want to contact me for any clarifications, or whatever, XXX Oh, one other thing ... I am new to this area and now have a new urologist, the top rated (by physicians in this state) urologist. I did not relaize how blessed I am. He said that with my RRP combined with the adjunctive radiation therapy, it was extremely rare for someone to be in my situation with regard to erections. He even asked me to attend the local PC Group, at least once, to let them know that "all is not necessarily lost", and I did just that.

27	First a comment section on dx was greyed out-- no typing accepted. How to put this in?? In early Feb, 2006, I experienced double vision while reading, with bilateral arm raises possible, hearing ok, no speech impediments or CNS signs. When same thing happened again the next day, I made an appt with opthamologist, who literally ran me out of his office after his MA took BP. I went to another clinic and asked them to take a BP and they reported 240/120. Follow up MRI revealed hydronephrosis, with lympodema on left leg. Left obdurator nodes obstructed, as was left ureter; left kidney was not working and collapsed. Bone scan lit up all along spine; bx showed GS of 8, and PSA of 50; pretty good for a first dx, eh? The second urologist made a big deal out of the high PSA, and blamed me for not "doing something," what was not stated. The 1997 PSA of 4.7 was not in my chart, and was done by a previous medical organization. A shock to be sure. Every doctor told what he could do---no one listed the range of choices, nor other specialities. No one talked of side effects or sexual problems, or incontinence, or scarred tissue, etc. Not to mention chronic fatigue, emotional liability, depression, sources of information, etc. Had not had sex in 10 years, so nothing was noted as "failing, or hindering." DX was acomplete surprise. Father died at 64 of unknown disease, was predecesed by his brother; grandmother brought heart disease into family, and had PTCA done several times in 1997, with ultimate LIMA, which helped immensly; went vegaterian, had always exercised, lost some weight (BMI was never >26), but worked at being heart healthy. No thought given to PCA. Had annual stress tests, met with internist regularly, but no PSA was done at any time to my knowledge.
28	First, thank you for doing this. The fight against prostate cancer needs all of our help. Ironically, prior to my biopsy, I was asked to participate in a XXXX Clinic research study of "false negative" PSA readings. At the time of my diagnosis I was at 1.2, low for a 50 year old.
29	Following surgery, I regained contenance in 2 weeks, and sexual activity with help of viagra in 6 weeks. At 3 months, no longer needed viagra. A couple of duplicate questions (maybe intentional?) that might give you some conflicting answers. Some questions seemed to be NA depending on how one answered the preceeding questions....so again, may impact the overall analysis slightly since NA was not a possible answer. In any case, hope this servey helps people, and look forward to seeing the overall results. Good luck!

30	FOR QUESTIONS 5-59 I USED THE TIME PRIOR TO SURGERY. IF IT IS POST SURGERY, ALL OF MY ANSWERS ARE WRONG. SORRY. AGAIN, THANKS FOR THE ABILITY TO COMPLETE THE SURVEY AND ALLOW ME TO MORE FULLY UNDERSTAND MY CONDITION. I WISH THAT I HAD TAKEN MORE TIME TO RESEARCH MY CANCER BEFORE I HAD MY SURGERY. I THINK I WOULD HAVE WISHED TO WAIT AND GATHER LOTS MORE INFORMATION BEFORE "GOING UNDER THE KNIFE". mY LIFE WILL NEVER BE THE SAME. I'VE SINCE LEARNED THAT SOME CANCERS ARE SLOW GROWING. I DID NOT KNOW IF MINE WERE. IF IT WERE "I SURELY WOULD HAVE WAITED".
31	from questions 55 to 59 is this your whole life or just the past few years? i answered as if it was my whole adult life. I have been taking Viagra for the last year and it seems to work good. Fairly firm erection for 70 yr old i still have the problem of dry ejaculation which i would prefer not to have. One urologist informed me that it was the drug xatrel (10mg) which caused my dry ejaculations.
32	Good luck with this survey, I hope you get useful information. I don't imagine mine is a huge contribution to the sum of PC/sexuality knowledge, it's probably a bit of a statistical outlier - my main reason for taking the troble I suppose. There again in a survey of 3000 I don't suppose you'll get too many 70-year-old, vasectomised, ex-smoking, reduced-drinking, acoustic-neuroma-excised, 7-year retired penetrative-sex participating, urethra-blocked, radical-prostaectomised cancer survivors!
33	Good survey, however some of the questions are too general. In my ciutation, some of the question do not apply. Several years ago, I caught my XXXXXXXXXXXXXXXXXXXXX and I called him on it. Well that caused a very big problem in my marriage. In the past X years my wife and myself have gone to 2 different marriage counsellors for help. Nothing was accomplished. The last counsellor told me "the marriage is over." I prepared myself, mentally, for the bad news and I am still living in that atmosphere. I am constantly subjected to mental abuse and consequentely my desire to be intimate has been destroyed. I don't dare seek outside affection because that is all the "world" needs to hear is that I have cheated. Then the mental abuse would be justified. It's sad that I can not give you better answers to your survey because I feel it is very important to other men and women. Please understand that many of the "sexual questions" I can't answer because I don't know. All I can say is befor I caught XXXXXXXXX, sex was good, very good. After that, in my wifes eyes I am a peice of Shit. I tried to get us help, but I am the outsider and you know blood is thicker than water. If my days of sex are gone, thank God that I have beautiful memories of the good old days. I really have no idea if the Prostate Cancer and it's treatment has had any effect on my ablility to have an erection, all I know is there is no need. That is unless things change and I doubt it. Thanks for listening, please help the other guys.

34	*Hard to answer since most of the 12 years were satisfactory and only recently started hormone therapy which changed everything. Other issue is that I am 12 years older so always wonder what would be the natural course of my potency if I was not treated or had Prostate Cancer. Once it metastasizes and survival becomes an issue one's oersoective on the importance of sex changes. I wish there were more resources available now to both my wife and i about achieving intimacy without intercourse. She does not like to use the internet for a source.*

35   *Have met men who simply gave up their sex life after PC treatment. Their partner was either indifferent or "tired" of sex and the man did not feel the effort to achieve erection worth it. So, they simply closed the door on that part of their life. Would the man and/or spouse feel differently with appropriate counseling about maintaining or re-achieving a healthy sex life? Don't know--and this survey did not explore how many just quit having sex, why they quit (i.e., was it PC related or simply age related and the PC was just the final straw?), was quitting what he/they really wanted, would they feel differently if they knew more about sexual function with age and/or PC treatment effects and methods to overcome problems, etc. (i.e., lubricants that easily combat vaginal dryness to eliminate painful intercourse; dealing with effects of hormone changes vs. libido; effectiveness of aids to achieving erections; etc....) I suggest understanding "how many" PC patients quit having sex and "why" are important insights into the whole PC and sexual function picture. Have also met numerous men who chose their PC treatment (or no treatment!) based on treatment aftereffects rather than treatment effectiveness for the PC. Most have felt their sex life would be so altered or negated by the PC treatment that they chose less effective treatments in order to avoid aftereffects. Their doctors seemed to take a neutral stance, giving alternatives and accepting whatever the patient elected to do. Doctors need the assistance of robust education programs to better inform PC patients and partners so that doctors do not have to be the sole advocate of preferred treatments...and PC patients need to get a better rounded insight--that the doctor may not have been qualified or comfortable providing--into PC treatment options and aftereffect countermeasures. Getting aftereffect treatment options from my own urologist was like "pulling teeth"--I had to be aggressively proactive to gain the procedures I now use for my bladder incontinence and ED. PC is life threatening and certainly life changing--you'll never get back to the pre-PC condition. Treatment needs to be a coordinated patient/partner/doctor effort and as aggressive as the situation will allow. The objective is to (completely) remove the PC as a life threat while preserving/restoring bodily and life functions/activities. Efforts, like this survey, should help support this comprehensive approach to PC and treatment.   Today some do get the info they need, but there are still many who don't. I meet men all the time who don't have a clue what their prostate does; they don't know why they need routine exams; they don't know why early detection is important--and that 40 yrs old is not too soon to start testing. (I've even been told their doctor downplayed PC all together as a concern before very late in life--this shouldn't be!)   Thank you for your work. Sex is a very important part of life. PC does not equal the end of a sex life--only a change in it...and, in some respects, my sex life is better now than before my PC diagnosis--go figure! We need to get the word out--PC does not have to be a killer and it doesn't have to stop anyone from living life!*

36	Hope this is helpful. I am 78 years old and lead a very active life. Untreated by one doctor he allowed my PSA to climb to 10.3. Have a very loving wife and give thanks to God every day for all my blessings. XXX XXXXXX
37	Hope this will help in the future - we (my wife & I) should have had better information prior to the surgery - not that it would have made a differance in having surgey - we could have been better prepared after - we had to research and discover for ourselves - it was difficult.
38	I am 67 and consider that I have a high Libido. Obtain gratification from movies most times. Partner is not missing sex but every few months we have oral and hand stimiluation that does nothing for me but provides organisms for her.
39	I am fortunite to not need any help I am not as active as I was before surgery, but I am cancer free for 5 years
40	I am now going to try erection injections. Nothing else worked so far. My first concern after surgery was incontinence. This whole nightmare was much more life changing that I was warned about. Little chance of incontience - had it , and then should be able to treat ED with drugs - no luck so far.
41	i am perhaps too soon after surgery to answer most questions - it's only been 3 weeks. Although my sexual desire is high, i have no erection capability. we are just at the stage to try viagra to see if that helps. we have not tried any other outside help to assist me in obtaining an erection.

42	I beleive that a lot more attention needs to be placed on studying the impact of radical prostatectomies on a man's orgasmic function. What really happens when the prostate and seminal vesicles are removed? What role do these organs (?) have in providing SENSATIONS to the brain which allow the male to experience an orgasm . . . with much less stimulation to the penis? We know that by massaging the prostate, a male can easily be brought to orgasm without even touching the penis. So why do famous surgeons (e.g. Dr. Patrick Walsh, etc.) think they can remove these organs without disturbing the orgasmic function in a male? We know that there is a distinct difference between ejaculation and orgasm. We also understand why men . . . after having RP surgery . . . will typically have dry ejaculations . . . except possibly for slight urine leakage or fluid from the cowpar glands. But when the sensations that are produced in the prostate and seminal vesicles are no longer there, the total amount of stimulation/senstion from various parts of the body have been reduced . . . perhaps significantly. In order for a male with RP to get over the orgasmic threshold, more stimulation is required from some other part of his body (e.g. physical stimulations . . . such as to the penis, visual stimulations . . . such as watching something erotic, thought stimulations . . . such as erotic fantacies, hearing stimulations . . . such as grownings from partner, touching stimulations . . . hands or mouth on partners breasts, partners proactiveness, etc. What gets me is that most articles or books I have read, tend to down play anorgasmia problems in men. I believe far more attention needs to be placed on this problem . . . especially with respect to men who have undergone RP surgery. I believe that the drug industry could develop drugs to help overcome the anorgasmic problems in both men and women . . . just as they did in helping to overcome ED . . . using oral medication. There are currently a number of over-the-counter lubercants intended to make gentials more sensitive to stimulation. These do help . . . but with more effort, I have no doubt that the drug industry could solve the problem. However, if medical community continues to downplay the problem and to write it off as a purely psychological problem, no one in the drug industry will ever invest in research to solve the problem . . . especially for men. Back in the 1950's, the medical community wroteoff ED as primarily a psychological problem. They were WRONG about ED THEN and they are WRONG about the cause of anorgasmia problems in men who have undergone RP treatment NOW.
43	I couldn't enter my test results on the first page; They were PSA 7.0 at diagnosis,7.5 at treatment, Geason 7 [3 + 4] Grade T2B,pot-op plus 2 months PSA 0

44	I did try lavita. When we have sex the first orgasm is very easy the second is not easy anymore. What typically happens is that after one orgasm for me and 2 for her we wait until the next day and we are good again. As you may have guessed she is a wonderful sex partner even after 57 years. The laita did not appear to help for the 2nd orgasm for me.
45	I do now have penile implant, and so do not worry about erections.
46	I experienced "erectile disfunction" (impotence-I prefer) prior to diagnosis of prostate cancer and used a vacuum pump to achieve erection. I always works. The first time I used oral aids, they worked the first time and thereafter nothing. I was not incontinent prior to surgery. Never recovered continence after surgery -- presently us a penile clamp -- works fine. Have very little trouble with incontinence while sleeping -- do not wear the clamp at night! Usually get up once a night. At 75, I'm not excited about using the pump as my wife (a hottie) isn't and never was pleased with the loss of penis warmth using the pump. I shall make an appointment with a male performance clinic soon -- recently established in XXX XXXX and see if penis injection will provide erection. If no erection, no cost!
47	I found injections worked very well for me. During hormone therapy my libido disappeared. Since being off treatment for the past year it has reappeared with a vengance. Mutual masturbation is semi-satisfactory. Since menopause and a hysterectomy, sex has been totally un-important to my wife. Bascially due to vaginal dryness and lack of desire.
48	I had a good sex life, although we had sex about four times a month, with no issues. AFTER SURGURY, it has been very difficult to be arroused, partially because it is so un-natural. everything has to be planned and to much effort to make sure everything work, that it is diffiicult to stay interested. Nothing can be spontaneous. It is easier to just ignor the need to have sex.
49	I had trouble giving my wife an orgasm with my penis when using penile injections. However, after the penile implant, almost always. I always make sure she gets an orgasm, if not with my penis, then using alternatives.

50	I have a lot of trust issues with the medical community in my city of residence. I prefer to hear the plain unvarnished truth and that is very difficult to obtain here! I have problems getting appointments for tests and consultations on a timely basis and have resorted to having some of them done outside of the country. If I had not taken the initiative I would have had surgery and the cancer would still have been growing in my bones. I had a urologist with a receptionist that lacked basic human relation skills. The oncologist that prescribed the hormone treatment had no interest in discussing control of the side effects. My perception of him is that he sees me as a 68 yr old man, not a 51 yr old man. I have been refused truthfull answers by staff of the cancer institute. I had a urologist renege on a promise of a bone scan after a few months. Current wait to get an appt with him is 4-6 weeks. I was always able to arrange same day appts to see a urologist in Mexico XXXXXX. Why are wait times so long for bone scans, MRI's Cat scans etc. when other countries will provide them in a day? In approx 20 tries to contact the receptionist of my oncologist I have NEVER had a human being answer the phone. Only a voice message! I am consistently being told less than the full truth. This includes the internet. Where for instance are all the side effects of anti-hormone treatment layed out? I believe much of the problem that I have with the medical community is that they are taught to "reassure" the patient. It is very frustrating to get them to speak frankly and truthfully. I have a good GP but it took 5 months to get an appt for a physical exam. Then more than 3 months to get a biopsy. Then 3 weeks to get results. I got the results on the way back to work in a foreign country and could not get an appt with the urologist for a follow up appt for 4 weeks! I could go on further in regards to the problems I have had with the medical community in my home city but it would be pointless. In short I am disgusted at the quality of care in an industrialized country like this. I do not completely trust any of them! Thank you.
51	I have been impotent since surgery. Tried the pump with no success as far as being able to maintain an erection. As my PSA never went to 0 post surgery I started hormonal blocking 5-6 months after the surgery and lost all desire for sex. During the of cycles I did regain some desire and we experimented with the pump but was not able to get an erection sufficient for penetration. We used oral sex and manual stimulation during the off cycles, though I must admit there has been no desire for sex on this last off cycle which ended on 1 Aug. 2009. My wife and I do hug, kiss, and embrace and try to mutually support each other.
52	I have been successful once with caverject injection. I tried numureous times but was unsuccessful in getting an erection or putting it in the right area and having swelling where I did not want it. I do not like to inject the needle and in Canada we cannot get the epi pen type device. My urologist showed me one time only and never asks how it is going since he showed me. So I have given up on getting and maintaining an erection.

53	I have everal problems that are very frustrating to wife and I besides those captured in the questions above: 1- pyronies disease gave me a curvature that makes masturbation difficult and penetrative sex very very difficult. 2-I experienced severe shrinkage of penis (length and girth) after surgery that makes intercourse difficult. On the rare occasions when I am able to penetrate, it isn't satisfying to my wife.
54	I have had an implant for 5 years, very satisfactory.
55	I have not had an erection sufficient to have penetrative sex since treatment.
56	I hope there are answers somewhere, to our problems
57	I hope you can differentiate the success of erections after the two main methods of radical prostatectomy. My current urologist (he follows up on prostate cancer patients after treatment elsewhere but does no treatments for prostate cancer on his own and is quite objective) says he has seen many patients after a radical prostatetectomy using the laporoscopic/robotic method and every single one cannot get an erection after treatment. I, instead have had an open retropubic radical prostatectomy with a world reknown expert surgeon and can now after 15 months since my surgery, get about an 80% erection without assistance and can get a full erection with the use of Viagra. Some mebers of the support groups told me all treatments were the same, or each person would say his was best. And I do not understand choosing the laporoscopic/robotic surgery justifying it on the basis of fast recovery. Because no matter whether you have the laporoscopic/robotic or the open surgery, you will be hanging around home for the next two weeks with that Foley catheter hanging out the end of your penis. Also I am completely urinary continent.
58	I know that I may not fit your criteria as I was always a Gay man, but I wanted to contribute. To help you classify me I provide the follwing information. I have always been out to my wife, but I never acted on my homosexuality until I recovered from EBRT/Lupron treatments in 2007. It was then important for me to understand who I might have been. Unfortunately, I waited too long and my Libido has been damaged by Lupron (3 - shots during the first year only). It does not look like it will ever come back, so it is difficult for me to find a place in the gay world. Fortunately, I am sexually happy with my wife. Thought it might be important for me to expalin this as to not hurt your data or help you interpret my answers.
59	I passed out when I self-injected. Then we went to Cialis with no positive result. Now back to injections, with my wife doing it. We have not had frequent sex beause it upsets her to do the injections and ruins the mood. Will keep trying.
60	I surely hope people fill this out and we can get to helping MEN with disfunctional problems. It seems more applicable with Radicals.
61	I tried to enter my birthdate but it would not accept the date! XXXXXX (man)

62	I tried to enter PSA numbers for question 10 but I kept getting an error message. It kept telling me to enter a positive number. I did and it still complained. My initial PDA was 2.4. My current PSA is 0.54. My staging is T2a, Gleason 6.
63	I tried to enter the stage for my PC at duagnosis but it was rejected. It was T2C
64	I want to make it VERY clear that all my attempts to have a reliable erection using injections or a pump were unsatisfactory, leading me to have a three-part implant at the XXXXX Clinic. Since then my sex life has been a total success for self and partner.
65	I was lucky enough to do the research on different treatment options available. After my urologist told me there were other ways of expressing affection to my wife, I realized I needed to be my own advocate and research and find the best treatment for me. He was sure that DaVinci Robotic Prostectemy would be the best option for me. When you are holding a hammer, everything looks like a nail. He was not truthful with me in explaining the options I had. I am sure he believes his treatment was the best option, but in reality it was not. I wanted the best quality of life post treatment. I am continually amazed that Proton Beam Therapy is not discussed more widely. I found it and am so grateful that it worked for me with very few side effects. Why this is not told to PCi patients is beyond me. My radiation vacation not only eliminated the cancer, but also taught me more about myself and priorities than I previously had known. I was treated at and the experience was nothing less than positive in every aspect. And this story is being told everyday by most of the patients being treated this way. If more men knew there was an option that was available that did not have the side effects associated with the treatment, I think many more would be spared the life changing things that happen with other treatments. No catheters, no drugs, no incontinence, no impotence, NO SHRINKAGE, no surgery, no recovery time, no blood banking, no leakage. I hope this helps, and perhaps your next survey will address Proton Beam Therapy.
66	I was very happy after having my surgery that sex became even more enjoyable then it was before having cancer. I was fortunate enough to have had a great doctor to perform the surgery and lucky that he was able to do nerve sparing surgery. After surgery I went on a regiment of viagra followed up with shots. Within 10 to 12 months I was able to get a healthy erection without shots. Today 4 years later I can get a full erection and enjoy sex. The one drawback has been that my penis is not as large as it was before having cancer. I would be willing to discuss in more detail the wonderful recovery i have had if it would help others.
67	I'm glad you are doing it. Hope it helps. I had no problems completing it.
68	In my situation I can reach orgasm with stimulation but cant get an erection for vaginal penetration.
69	Intercourse attempted, but never successfully since surgery.

70	Intercourse has never been important to my spouse. My cancer has not changed that.
71	It has been almost 9 years since my surgery, and it has been very depressing for me. I don't feel the doctors realize the impact it has on patients, and they just want to go on with their surgery and let the patients struggle on their on; they don't want to hear about it. (At least the ones I have delt with.) Don't get me wrong, I'm glad to be rid of the cancer, but it has sure left with a price; I sometimes wonder (at my age) if I would have been better off on the "watch and wait".
72	It is my personal opinion that very little realistic information is available to the prostate cancer patient and his partner. This is truly a disease that impacts the couple. I wish that I had known about Trimix after my surgery. I would have started that within 2-3 months after my surgery. I truly believe that this would have given me an opportunity to heal the traumatized nerves. It would have also helped me avoid the degree of depression that I experienced. Every man who comes to me now, I advise them to demand Tri-mix or Bi-mix from their Urologist....or change their Urologist. Also, this is a couple's disease....more counsel should be provided to the couple. Finally, why is so little information about Prostate Cancer given in the media and by our government. Breast Cancer has mass coverage in comparison. There is definitely a gender bias in the media and government. Men are just advised to deal with it and keep it secret.
73	It seemed to me that the survey presumes all the sexual difficulties are a result of prostate cancer treatment. In my case, my ability to get and maintain an erection seems largely tied to libido. It now takes more to stimulate me than it used to, but when well stimulated I'm able to get and maintain an erection just as well as before my prostate treatment. The drop in libido (and drop in erections without stimulation) began to occur about 8 years after my seed implant.
74	It's depressing.
75	Just to re iterate. Not all questions distinguish between now (on hormone therapy) and a year ago, and a year hence, when I wasn't and won't be
76	last part of survey contains ambiguous language.....does 'intercourse' include mutual masturbation and oral sex........ ie, what 'is' is
77	Libido seems to have decreased and spontaneous erections are harder to achieve and maintain
78	long over due
79	MD never discussed penile rehabilitation, just "give it time"... and then it was too late. Wife objected to injections and I'm too old for implant. I do not trust MD's at this point in my life.
80	More men who have nerve sparing surgery should use the shots.
81	My answers are based on not having a partner (wife passed away 2years ago) & having my surgery 3 weeks ago
82	My doctor had me on a 1/4 viagra pill in the first three months after surgery. Once I saw I was getting erections, at about 9-12 months post-op, I resumed sex.

83	My only issue is no ejaculation. There was a period of time for about 2 months that I could not get an erection. I spoke with my radiologists and he referred to a clinic in XXXXX but could not get in. My biggest issue is no one wants to talk about this subject especially doctors. No discussion prior to treatment so that my wife and I could be prepared. Another side effect that I had during the first year after treatment was pressure and pain in my rectum when I had an erection.
84	My partner and I were not terribly sexually active prior to my being diagnosed with prostate cancer. Having prostate cancer and being treated for it simply eliminated sexual activity from our lives. I am sad about that, but I don't know what to do about it. My partner doesn't want to discuss the matter, I have to take drugs (Lupron, etc.) that pretty well eliminate my sexual abilities, and the doctors I have dealt with have been utterly useless on the subject. It is like they simply don't care about how the disease and treatments for it affect you mentally. If whatever they do keeps your PSA low, as far as they are concerned they've done their part, and how much damage has been done to you/your marriage in the process is of no consequence. On common theme in the support group I attended for a year or so after being diagnosed/first treated was the phenomenon of what we called "The Incredible Shrinking Penis": --It didn't matter what treatment people had had, whether surgery or radiation, everyone in the group was as certain as they could be that their penis was 30 to 50% smaller afterwards. I KNOW mine is. But universally doctors say it isn't true/doesn't happen/etc. They lie. It does happen, and it does matter. Good luck trying to shed some light on this subject. It is huge, especially for couples in their 40's or 50's who ideally should have had many years of sexual intimacy to look forward to. It is pretty bleak when one is this age, and sexuality is dead. Damn doctors need to get their heads out of their asses and start dealing with this issue more honestly and openly, and not leave it entirely to the patient to try to figure out on their own. It is as much a part of the disease as is any other part.
85	My partner is passive during sex. Leaves it to me. Without stimulation I have a difficult time at 78 years of age. I believe that I would not have much of a problem if I had not had prostate surgery.
86	My prostate cancer was all ready in stage 4 when found and spread to my bladder, seminal vesicles, vas deferens, neurovascular bundles on both sides, lymph nodes on both sides of abdomen, and soft body tissue. My prostate gland was over 90% prostate cancer. Even with positive lymph nodes they operated for the purpose of debulking and to help with my survival time. With all that was removed and also taking Lupron for over 11 years straight with no break, my libido was basically zero and then to lose all of my posessions and savings, insurance topping out while a patient at MD XXXXXX under a protocol my libido was totally gone and depression was at it's worst.
87	My sexual connection with my love is mostly "carinos": caresses, touching, oral sex from me to her, aimed at clitoral orgasm for her, laughter and kisses, hugs, etc etc....

88	My sexual lilfe is nothing after surgery. After several failed attempts my partner has completely given up and lost all confidence and interest in me. With the exception of a few self masturbations, there has been no sex in my life. ps. my wife is not interested enough to take this survey.
89	My wife achieves sexual satisfaction from my touch over 90% of the time prior to my attempts to have intercourse with my semi- or erect penis.
90	My wife and I pleasure each other orally and that is fine with both of us.
91	My wife and I think you are covering a very important area that has been ignored by the medical community...Thank you for your research and we hope you can reach many of those survivors who think their sexual life is now over after prostate surgery....
92	My wife died about 18 months before treatment started so sexual opportunities have been limited. I can ejaculate with a partial erection which I can achieve without medical help. My orgasms are not dry but volume has decreased.
93	My wife has gone through menopause and has experienced vaginal dryness for several years. This has caused vaginal pain during intercourse and consequently she has pretty much shut down. I keep trying and suggesting alternate methods but have had no success to date. I'm glad that someone is studying this and hope that there will be some suggestions based on the results. I am very interested in the results and hope that they will be published when complete.
94	My wife left me just before I was diagnosed which makes my recovery quite difficult. It's hard to put on your eHarmony bio that you have ED. One woman I dated seemed interested until I told her I had recent prostate surgery and she never saw me again. Very discouraging without any form of helpful mate. Wish there was a group of caring women who would aid the victims of this emasculating disease.
95	My wife was not willing to give me the opertunity to attempt intercourse and I had no sense of touch or feeling after the operation. Testosterone levels were very low and therefore my own sexual drive was reduced to very low. Antiandrogen drugs given prior to the operation lowered Testosterone to very low levels and I never recovered.
96	Never tried assistance. To date it has not been necessary.
97	Nice range of questions and validation attempts.
98	nnot enter TIC in stage
99	None

100	Obtaining an erection has not been difficult. Having intercourse with a 4" penis is difficult under any circumstance, but add a wife without any seeming interest, and it isn't attempted. Before hormone therapy, I may have been 5", but that seemed to be enough. Our "love making" before prostate cancer had involved lots of manual stimulation to my wife which brought multiple orgasims. The time of penetration was relatively short, but also seemed to produce orgasims. You probably should not count my responses to this survey since I don't have ED. At least the survey allowed me to get some frustrations off my chest. Oh yes, I have talked with my wife a little about our lack of love. She doesn't know what to do and doesn't seem to be interested much in my feelings of inadequacy. She hardly ever was the one to start love making before. It was almost always up to me. She's doesn't have a clue now.
101	Once again, this survey does not apply to losers like me
102	Only treatment I have tried to date are the oral medications I stated plus Muse. None worked . I must say Cialis gave me the most "promising" signs as I was able to have an erection whilst masturbating a couple times .. I will be trying the injections in January 2010 as I am part of the Prostate Rehab program run out of XXXXXXX Hospital.
103	Page 2 does not define whether the questions are before or after cancer treatment. I answered as for after surgery.
104	Penetration is not the problem, not reaching orgasm is. Information on this problem is in short supply.
105	Penile shots are what we decided on way up front. They worked quite well for us. After reoccurrence, and three intermittant hormone treatment periods, problem now is libido is non-existent. I have no urge to even start any floor play, etc. (got off zoladex mid Dec 08, now mid Oct 09; maybe body is just not going to "snap" back. Snap back is a bit of a joke!)
106	Question around with med aids or assistance too general - eg drugs dont work - injections 100% but painful so not completely satisfactory.
107	Questions 61-63: are answered in the past tense: which is to say pre prostatectomy.
108	Restrictive format. Married for 9 years after treatment to supportive but increasingly uninterested spouse; sex infrequent at end. She asked for divorce -- "done with the marriage thing." New loving partner of 3 months -- not only deeply loving, but sex is a usual daily part of the relationship, without asking or begging or scheming or manipulating. How come I had to wait until I'm 65 to discover this woman??
109	Since my operation, I have not had to use any medical aids or assistance.
110	Since viagra works so well, I really do not attempt intercorse without it. My wife is able to acheive orgasim with use of vibrator & lovemaking.

111	Some hard to answer. Had a partner 1st 3 yrs. Dated sporatically past 7 yrs, only 1 relationship lasted several months. Intimacy continued satisfactorily only with 1st partner. Extremely difficult to start new relationships with near zero libido. Often doesn't seem worth the trouble.
112	Some of the implications of these questions appear to center on whether of not we can have successful intercourse or not. Many men I believe must reach a point where they come to peace with the situation. Is sex more important than life. We also must learn that our partner, especially nearing age 60 is more interested in feeling loved, than having rousing sexual encounters. The intimacy that results when the man concentrates on the female's needs, and bringing her to an aroused state and orgasm is a satisfying sexual encounter for both.
113	Staging with integer does not apply in Canada: Revise item 2
114	survey does not cover use of drugs like inj lupon
115	Thank you very much. I would be interested in the survey results when your study is completed. I'll send a follow-up email. I feel that my prostate cancer situation may not be exactly what you were looking for....since I have not had any invasive treatments. I am in an active surveillance mode. My cancer was found incedently after a TURP procedure. My digital rectal exams have always been normal and my PSA has been flat (around 1-2) for the 10 years I've been getting them. A follow-up normal biopsy (18 cores) found no cancer. I've thus had no changes in my sexual function before and after diagnosis. Good luck.
116	Thanks for being willing to research this area. For individuals like me, it's important to see that there are others out there that have similar issues.
117	Thanks for doing this. I wish that the pills were cheaper so I could afford them.
118	Thanks for looking into this.
119	Thanks for the concern, I my case and I am sure in others, injecting medicin into the penis is somewhat painful and difficult to do. A instrument to assist in this proceedure would be great.
120	Thanks for the opporunity. While I have a spouse, I have no sexual partner. She has reached the conclusions that no sex is better then flusterated sex. She has given up and refuses to even participate in this survey.
121	Thanks, I wish I had a partner willing to work with me.
122	The frequency of natural erections seems to be diminisioning the past 6 months but I have also moved 3 times in the past year and renovate for the last 6 months so I have been over tired a lot of time which I know greatly effects quality sexual timing.
123	The main issue is achieving an erection sufficient for penetration. Once penetration is achieved, the erection seems to improve and most of the time intercourse is satisfactory

124	The only medical assistance that worked was my penile implant. Sex has changed with the implant - arousal is gone - has changed the relationship with my wife - something sweet has been gained. Still trying to define that.
125	The problem with this survey is that it does not allow for stages of treatment - ie it assumes that repondant has only one form of treatment. ie. surgery followed by homone deprivation several years later is not real catered for. Therefore some of the reponses may appear to invalid
126	The range of time mattered a lot. It's been over two years. The first six months there was almost nothing; now after two years we can often have enjoyable sex without Viagra, although we usually do it with Viagra to be sure we have the fun we want.
127	The requested numerical number of the stage of cancer which was T1c was not allowed after many attempts to "correct" the "positive number" I resported to using just the number 1 and it was allowed. Seems not to be the best information for the purpose of the survey.
128	The survey seems quite complete exept for two areas. The smoking and drinking questions did not give enough options to give correct answers. The survey was not accepted because the PSA, Gleason, and Stage was not it whole numbers. I put in whole numbers several times and it still would not accept it. Finally I removed them.
129	The survey was good. I am married for 42 years and my wife does not appreciate the importantance of a sexual relationship - therefore we have not had intercourse or any form of sexual relations in over 15 years.
130	the wording on the last 2 series of questions is confusing. Are these questions about conditions following treatment? (or prior to treatment?) Also, the number of cigarettes per day in that series of questions is confusing (number of packs per day???
131	This last block of questions (Without Medical Aids/With Medical Aids) is not really clear. I am assuming that this last "With Medical Aid" means since I've had my penile implant???
132	This survey assumes that there has been some significant amount of time since prostate cancer treatment. In my case, it has only been one month so it was hard to answer some of the questions.
133	This survey is so slanted toward straight couples that I almost didn't complete it.

134	This survey seems to assume the survivor is heterosexual. Because I have been bisexual all my life and the father of 2 children I raised from childhood due to their mothers' death from breast cancer, I have been "closeted" since my wife's death. My wife was aware of my sexuality - it was never an issue - we were happily married. My son is married, my daughter, who lives with me - ironically "came out" to me 3 wks. prior to my diagnosis - so naturally I came out to both my adult children at this time they are ages: XX and XX. I was ready to go out and take my chances at a relationship when this occurred. Since puberty until 6 months after treatment I could count the days of "no ejaculation" on one hand and then it was only due to hospitalization, or because I was in a place where it would have been difficult to "assume position" - naturally because of the frequency masturbation has been my daily routine - still is even without erection - the loss of erectile function has been emotionally devastating to me, as well as the shrinkage of the penis due to treatment - I never knew that at age 60 my sexual life and hope for a partner and/or intimacy would be over - I've become embittered and dealing with suicidal depression since.
135	This survey was informative in and of itself. The questions have informed me that others are experiencing what I am.
136	Though not currently sexually active I do have regular nocturnal erections and do masturbate. Orgasms are a bit more painful since radiation, but not enough to be discouraging. As result of the surgery I also have a slight curvature of the penis (doctor diagnosed). The answers on the previous page (erections/intercourse) were from my 2nd marriage. Part of the reason for the divorce in my first marriage was a lack of sexual intimacy. My second wife was quite different from the first. She thought the lack of ejaculate was wonderful, made oral sex much more pleasant. The curve in the penis was another plus, she felt that it stimulated her in the right place. Good survey, but I think that I am a little off the curve. Hope it helps.
137	Thought provoking and timely...
138	too long....
139	TriMix injections work but just the idea of injecting the penis is a turnoff. At 76 years, with 4 marriages (and divorces) and 50 plus previous satisfactory relationships, the need for intercourse is not high on my list - although my 70 year old love would like it.
140	Very interesting survey. If I can help in anyway you can reach me at this email adress or

141	We are currently not having sex...although I have a penile implant which works...I still have performance failure issues and struggle with my inability to ejaculate (sometimes urinate) so our sex life is not so good now...it scares me...as I also have PTSD issues which have resurfaced as this is tied to agent orange exposure in Vietnam ...I do not think my wife is having as much of a struggle with this as I am but we try to stay connected emotionally ..thanks for helping us...no one really prepares prostate cancer survivors for these issues...but our US TOO SUPPORT group has been a REAL BLESSING!
142	We assume that "with medical aids or assistance" includes medication such Viagra.
143	We discovered that it is not necessary to have intercourse to achieve orgasm.
144	We tend to keep too busy and have not been concentrating on sex lately. Levitra almost works.
145	We use a pump almost entirely, and the answers here are based on it's use.
146	Whats the difference between an egg, an whipped cream and a blow job. Answer You can beat an egg, and you can beat whipped cream, but you can't beat a blow job!
147	Will help future survivors more than me. Covered lots of areas I no longer think about anymore.
148	With penile shots it worked reasonably well. I had a penile implant, but, had complications after surgery and it was removed. Two months later I have not tried shots again, but, am not optimistic it will work.
149	Without giving away who I am (maybe too late), over the last few years (since my Radical Prostatectomy), I have been an advocate for a unanimous patient driven survey. Your questions are very good. One thing that is missing is a "Yes" or "No" category, in some places. This leads one to believe that I didn't want to answer the question. When to do so, would be misleading. For instance - 52. Do you have an informal support network? 41. If your physician discussed with you the possibilities of having problems with sexual function after treatment – he did but I can't remember the percentage. 43. If you are on androgen deprivation - I was on ADT but am not, at this time (hopefully never again). Yet the questions around this are for those presently on ADT. As it wasn't that long ago, I could have provided helpful answers. I wish you the best of luck.

33% of the survivors took the time to comment. These comments have made this survey worthwhile and we thank you for your honesty and openness. We will make every effort to make sure that you are heard and helped.

We cannot overemphasize how helpful these comments have already been for many survivors and their partners.

We will do our utmost to honor your efforts and get this information out to those who will benefit from it. We hope that this information will continue to be a resource for everyone dealing with prostate cancer.

Prostate Cancer Survivor PARTNER Sexuality Survey

Introduction to Actual Survey

A Consent Form for this Study is an agreement to do the survey dealing with the after, effects of prostate cancer treatment on the sexuality of both partners and to accept all that it encompasses.

We invite you to take part in a research study being conducted by Dr. Joel Funk, Dr. Jo-an Baldwin Peters and Court Brooker. Your participation in this study is voluntary and you may withdraw at any time before the survey is completed. The study is described below and identifies any risks, inconveniences or discomfort that you might experience.

Participating in the study may not benefit you directly, but we expect to learn things that will benefit others. You should discuss any questions you have about this study with Dr. Jo-an Baldwin Peters who can be contacted at dr.jo.an.p@gmail.com.

Purpose of the Study

Through this study, we anticipate gaining more knowledge about how the after effects of prostate cancer treatment impinge on your sexuality and that of a partner. Recommendations can then be made to treating physicians and support groups, and appropriate materials can be developed in relevant formats and distributed in accessible outlets to prostate cancer patients.

Study Design

This study has been reviewed for content, ethical and confidentiality issues by a panel of experts. The survey invitation will be posted on several related websites for three to six months. All submissions will be completely anonymous. At no point will your name or other identifying information be requested or acquired. It is our intent to halt the survey when 3,000 responses are received.

Who Can Participate in the Study

You and your partner can participate in this study survey by visiting several websites or receiving an email link. The survey can be accessed/shared by copying the following links and pasting the appropriate one into your browser address field or emailing the link to other known prostate cancer survivors:

Survivor Survey Link

Partner Survey Link" No longer operational Survey ended November 2011.

Prostate cancer survivors without a partner may also participate and are encouraged to do so. As this survey focuses on men who have received treatment for prostate cancer and their partners, both partners are ideal candidates to participate in this survey. Prostate cancer survivors without a partner may also participate and are encouraged to do so.

Who Will Conduct the Research

Dr. Joel Funk is the Principal Investigator of this study together with Dr. Joan Baldwin Peters (PhD) and Court Brooker.

Dr. Joel Funk is a certified Urologist associated with the Yavapai Regional Medical Center in Prescott, Arizona, and Assistant Professor of Surgery at the University of Arizona in Tucson, Arizona.

Dr. Baldwin Peters is an independent researcher and the wife of a prostate cancer survivor.

Court Brooker is a prostate cancer survivor experienced in communications and media.

The study design was a collaborative effort between Dr. Joel Funk , Dr. Joan Baldwin Peters and Court Brooker.

The survey will be administered via the internet.

What Will You Be Asked To Do

The following survey consists of a maximum of 45 questions. Based on a number of trials it is estimated that the will take you 15 to 25 minutes to complete. This is the only task that you will be asked to do. Please note that detailed personal questions will be asked regarding your past and current sexual ability as well as your sexual orientation and MUST BE COMPLETED INDEPENDENTLY BY EACH PARTNER.

Possible Risks and Discomforts

You may experience some emotional discomfort while answering the survey as many of the questions are specifically about your past and current sexual ability. You may skip questions you find too difficult to answer. You may also withdraw from the study at any time.

We will make all publications based on this data available through our website by posting information about how to obtain them as soon as they are published. We will provide copies in PDF format to all who request them.

There are no direct benefits to you for participating in this study. However, our study may help us identify areas where better information is needed and in what format and for whom.

Compensation / Reimbursement

You will not incur any expenses beyond the time it takes you to complete the questionnaire. Therefore, no compensation is being offered.

Confidentiality and Anonymity

The data collected by this survey will be of a personal and sensitive nature, but will not be linked to any identifying information.

All data will be retained on personal computers for approximately two years under password-protected files, at which point the files will be erased.

Questions or Concerns

Any questions or concerns about the study can be addressed to Dr. Jo-an Baldwin Peters at Dr.Jo.An.P@gmail.com. It should be noted that in contacting her by email you will be providing her with an email address that may compromise your anonymity. However, your email address is not linked to the questionnaire in anyway.

Therefore, Dr. Baldwin Peters will not be able to connect you with your survey responses unless you provide her with sufficient information to do so.

If you are blocked from submitting or advancing whilst completing this survey it may be necessary to revise or re-enter some data. The system will return you to the question or questions requiring revision. Instructions will appear in red and be marked with an asterisk.

Unfortunately this system disallows fractions and requires dates to be fully completed. Correcting the data will then allow you to submit your survey.

Survey Questions and Results

1. Consent form for the study to identify how the after effects of prostate cancer treatments impinge on the sexuality of individuals and their partners. I have read the explanation about this study. I have been given the opportunity to discuss it with Dr. Joan Baldwin Peters and any questions have been answered to my satisfaction. I hereby consent to take part in this study. However, I realize that my participation is voluntary and that I am free to skip questions and/or withdraw from the study at any time. I also authorize the use of quotations from any of my answers to demonstrate my individual sentiments so long as I am not identifiable. By clicking on the 'agreement' button I electronically sign that I accept the terms presented above. (answer required to continue survey)

Answer Options	Response Percent	Response Count
I agree	100.0%	193
I disagree	0.0%	0
	answered question	193
	skipped question	8

Of all the survivors who had partners, only 43% completed the survey. This was a pity as it was a missed opportunity for their voices to be heard and counted.

2. Birth dates (your partner's birth date must be filled in) Example: For 14th December, 1930 enter DD=14 MM=12 YYYY=1930

Answer Options	Response Percent	Response Count
Your birth date	97.9%	189
Partner's birth date	99.0%	191
	answered question	193
	skipped question	8

Although there was an opportunity to match 193 partners, we were only able to match 68 couples. There was confusion between using the European/International method in completing the date of birth. As this was one of the rare surveys that included partners, this was disappointing.

3. I currently reside in

Answer Options	Response Percent	Response Count
Current City/Town	95.8%	182
Current Province/State/Other	98.4%	187
Current Country	96.3%	183
answered question		190
skipped question		11

This data needs to be broken down by country.

4. Relationship?

Answer Options	Response Percent	Response Count
Have no partner	1.1%	2
Have a partner	98.9%	186
If you have a partner, how many years have you been partners?		185
answered question		188
skipped question		13

Partnership duration ranged from 4 months to 63 years.

5. Your education?

Answer Options	Response Percent	Response Count
High School	29.7%	57
College/University graduate	42.7%	82
Post gradauate	21.9%	42
Other	5.7%	11
answered question		192
skipped question		9

The partner respondents were highly educated, which in itself creates a bias.

6. Your race?

Answer Options	Response Percent	Response Count
Caucasian	94.2%	179
Asiatic	0.5%	1
African American	2.6%	5
First Nations	0.5%	1
Other	2.1%	4
answered question		190
skipped question		11

It would appear that African American partners are not well

represented in this survey. The predominance of Caucasian partners creates a bias.

7. The following would best describe me

Answer Options	Response Percent	Response Count
Heterosexual	97.9%	185
Homosexual	1.6%	3
Bisexual	0.5%	1
Other	0.0%	0
	answered question	189
	skipped question	12

Inadvertently, the question format made it difficult for homosexual partners to complete some of the questions. We apologize for this oversight.

8. What type of treatment/treatments did your partner undergo? (can check more than one)

Answer Options	Response Percent	Response Count
Don't know	0.5%	1
Cryosurgery/Crotherapy	2.6%	5
Radical prostatectomy	30.2%	58
Radical retropupic prostatectomy	4.7%	9
Radical nerve sparing prostatectomy	25.5%	49
Robotic prostatectomy	22.9%	44
External beam radiation	17.2%	33
Proton beam radiation	1.0%	2
Intensity Modulated Radiation(IMRT)	6.3%	12
3 dimensional conformal radiation therapy (3D-CRT)	3.6%	7
Brachytherapy/seed implants	8.9%	17
High dose rate Brachytherapy	0.5%	1
High intensity focused ultrasound (HIFU)	0.5%	1
Hormonal drug therapy	33.3%	64
Chemotherapy	3.6%	7
Active Surveillance	3.6%	7
None	0.5%	1
Other	3.1%	6
Comment		30
	answered question	192
	skipped question	9

Over 50% of the men had some form of prostatectomy. The survey was designed to identify post-treatment sexuality issues and as this is a major problem for prostatectomy patients it could create a bias. If this is the case, it is unfortunate as sexuality issues arise to a greater or lesser degree, either early or later on, no matter what the treatment modality.

Q8. Respondent Comments

1	005 Radical prostatectomy, 7 wks radiation, 2 yrs Lupron; 2010 5 wks radiation & Lupron until 2 - 4mo PSA's result in ZERO (CANCER TWO TIMES)
2	cancer cells were incidentally found during a routine screening after my husband had surgery for an enlarged prostate. He had urinary incontinence after the surgery, which improved dramatically after 6 months, although isn't completely back to normal. This unforseen effect impacted our sex life for a while, but is now back to normal and is unrelated to the cancer, although a precurser to finding it.
3	Cavermap was used during the surgery.
4	Cryosurgery on March 22, 2010
5	Currently, my partner is in a clinical trial. He is taking NDGA.
6	davinci robotic nerve sparing prostatectomy
7	Diet change and many supplements and pharmaceuticals
8	Due to concerns over incontinence and impotence, partner has been reluctant to seek treatment. Is considering radiation.
9	Followed by 6 month PSA readings
10	He experienced a complicated post operative recovery, his anantomosis did not seal well and he had an indwelling foley catheter for 2 months. He had several cystoscopies and the contrast media (and urine) leaked into his pelvic cavity and caused significant inflammation
11	He is now on the abiraterone drug trial
12	He is on his last hormone shot and expected to regain testosterone in May.
13	He was diagnosed a year ago, had the prostate removed early this year and has just completed his course of IMRT and is still having hormone therapy
14	I don't exactly know the kind of treatment my partner did undergo. I would say radical prostatectomy but I'm not sure.
15	Is also schedule for radiation treatment in a few months.
16	laparoscopic nerve sparing radical prostatectomy
17	leukine, high-dose ketoconazole, cortisone, estrogen patches
18	My husband did extensive research before choosing robotic surgery; it went very well and his recovery time was minimal.
19	My husband was diagnosed 10/2002. He passed away 9/2009. He had many different treatments as his disease progressed. Each treatment impacted our sexuality until we reached a point when it became too difficult to achieve penetration and orgasm. We grieved the loss of our sexual intimacy. We continued to hug, hold hands, caress one another and enjoyed tremendous satisfaction just being next to one another.

20	My partner Larry had had a RP in 2001. We met in 2004 and were together until he died in Nov. 2006.
21	No sex since 1st cryotherapy on 6-05-1992.
22	One nerve spared; the other was removed
23	Ongoing
24	Radiation was first, then salvage surgery.
25	This has occurred over many years, I am answering as to the last course of treatment only, IMRT and hormone therapy for 9 months.
26	Was on hormonal drug therapy for 10 years before radiation.
27	We had to make a choice about what form of treatment would be best. The surgery was done almost immediately as there was bilateral cancer within the prostate. I would say this is one of the most stressful times I've experienced.
28	We're waiting on referrals from the urologist to the oncologist.
29	With the removal of his prostate, we had no idea, his penis length would reduced by the size of his prostate, which was enlarged. He went from 6" to 4" in length. His girth went down, too.
30	ZOLODEX RADIATION THERAPY 40 SESSIONS CASODEX VIAGRA

9. How often did you have penetrative sex before prostate cancer treatment?

Answer Options	Not at all	Less than once a week	1 or 2 times a week	3 times or more a week	Response Count
Before treatment	6	59	79	37	181
After treatment	78	50	27	8	163
				answered question	188
				skipped question	13

Penetrative sex decreased dramatically after treatment. Hopefully, alternative methods were used to maintain sexuality.

10. Were you innovative in your sexual practices (did not only use the missionary position)?

Answer Options	Traditional	Innovative	Response Count
Before treatment	41	134	175
After treatment	43	94	137
		answered question	182
		skipped question	19

It is postulated that innovative methods help maintain sexuality, especially if practiced prior to treatment.

11. How would you rate your partner's ability to have a spontaneous erection after treatment for prostate cancer?

Answer Options	Never	Sometimes	Always	Response Count
Within 3 months	126	39	4	169
Within 6 months	108	41	7	156
Within 12 months	87	40	14	141
Within 18 months	72	39	13	124
Within 24 months	65	36	11	112
Longer than 24 months	62	28	18	108
Comment				63
			answered question	179
			skipped question	22

The answers indicate that there is sexual recovery after prostatectomy but it often takes time. It is important to maintain penile tissue by using medication combined with either penile shots or pumps (1.2.3). There is also some indication that with radiation, sexual deterioration occurs over time. This could be age-related and not necessarily treatment-related.

Partners: we thank you for clarifying this tough, extremely personal issue. All your comments and hints are an invaluable source of help for others having to deal with these very difficult and challenging issues.

Q11. Respondent Comments

1	Only been 3 months since surgery
2	After 4 years, his cancer returned and he is back on hormone therapy. Erections occur sometimes
3	After salvage surgery, there was no hope for intercourse. Nothing was possible.
4	Answered question as frome date of prostate removal
5	Biochemical failure. Hormonal treatment at 4 years after brachytherapy.
6	Cancer returned after aprox 2 yrs & Lupron ejections diminished spontaneous erections
7	Dx 10/2002 w/aggressive disease w/bone mets. HDT & Chemo 1st 14 months. Radical prostatectomy 02/2004. As disease & treatments progressed, ability to have erections decreased, interest in sexual intercourse decreased (HDT) & incontinence increased. By 2007 unable to have successful erections. Husband passed away 09/2009.
8	erection not firm
9	Erection was poor and not sustained after seeding
10	Erections not hard enough for intercourse sex
11	Has only been six months since surgery.

12	He could not maintain an erection and sometimes had to deal with unwanted urine.
13	He did not have nerve sparing surgery.. uses Bi-mix injections
14	He had penile implant after about a year.
15	He had treatment 4 years ago. We have been together only 2 years
16	He is only 18 months from his surgery.
17	He must use daily Cialis for "Always"
18	He still does not have a spontaneous erection.
19	He was having difficulties just prior to diagnosis and has never regained the ability to have an erection of any form.
20	His treatment was minimal. He refuses to try to have any intimate connection, physical or emotional with me.
21	Hormone therapy was not instituted until about 24 months into treatment.
22	Husband believes he has an erection suitable for penetration, when in reality he does not.
23	I did not become intimate until after 3 years after his surgery. He furnished me with this prior knowledge.
24	I don't know how you define spontaneous, but when stimulated, he can get an erection.
25	I met him several years after surgery so the initial responses are based on what he has told me
26	I never regained potentcy-I had a penile implant 9 months ago
27	I would describe the erection as only slight.
28	It has been 18 months since his surgery with no sign of recovery though the surgeon assured him there is no reason not to expect recovery.
29	it has been difficult
30	It has not been 24 mos yet
31	It has only been 4 months since his surgery
32	It has taken 2 years to recover a 'normal' testosterone level after 12 month's worth of Zoladex implants
33	It hasn't been a year.
34	Larry could only have an erection using caverject.
35	My partner had had his treatment at least 4 years ago when I met him.
36	Needs medication. Surgery was in May 2009.
37	Nerves were damaged and could not get erection easily without alot of hand massaging or oral sex and then not very good. Got boners during night sleep - didn't last
38	NEVER
39	Never a full erection. Partial erection with some type of medication and then isn't full penetration
40	Never the same after treatment. Always requires stimulation, not hard enough for penetration. We have oral sex only

41	No spontaneity whatsoever. He tried Viagara, which gave him a headache and nasal congestion. He tried injections, which eventually worked once he got the right dosage, but then it was taken off the US market. The last injectable drug was a powder that had to be mixed and then injected. Nothing was ever like it was pre-surgery....he had no interest, no desire, and hated playing scientist with mixtures and injections.
42	Not a full erection anymore
43	Nothing worked well and my husband was not at all interested once he was on HDT/CAB
44	One year and one month since surgery.
45	Only eight months after the surgery
46	radiation was first in 1988 - sex was ok cryotherapy was second - no sex after. cryotherapy.
47	since he has opted for active surveillance, there has been no change in his ability to have an erection
48	Some impairment without treatment. Viagra helps some, but maintenance of erection is a problem.
49	spontenaity was gone
50	still early ..do not know....
51	Still on adt therapy for another 6 or so months, awaiting further developments
52	surgery 8/11/08
53	Surgery has been 6 months ago.
54	surgery was 8/07
55	The surgery removed all ability, the hormone treatment removed all desire and essentially my spouse's genitals.
56	The surgery was only in September. I have no desire for sex now as it has "messed up" Fortunately we are good freinds and that's still awesome.
57	There has been NO erections since 2005
58	We are currently three months post treatment.
59	We are still intimate at least twice a week. We are using a vacurect and viagra. Neither are really helping. We still TRY to have penetrative sex.
60	We have only been together for 3 years - my partner had treatment 9 years ago and is still going intermittant hormonal treatment with Casodex
61	We have tried many "helps" including injections.
62	we only been married 4 years and he had the treatment in 1991.
63	While he has no difficulty obtaining an erection, he has great difficulty maintaining one. He also lost his ability to control his orgasm. He now requires direct stimulation to maintain erection and to achieve orgasm. Three strokes and he's gone.

12. What things were tried to regain an erection? And in what order?								
Answer Options	Tried 1st	Tried 2nd	Tried 3rd	Tried 4th	Tried 5th	Tried 6th	Tried 7th	Response Count
Nothing	28	0	1	1	0	1	1	32
Self masturbation	26	14	9	2	3	0	0	54
Mutual masturbation	32	47	21	5	3	0	1	109
Oral sex	13	28	23	16	3	1	2	86
Sex toys such as vibrators	2	3	10	10	5	0	0	30
Sex therapist	1	2	1	0	0	0	0	4
Erotic magazines	0	0	1	0	1	0	0	2
Erotic movies	1	2	4	5	5	2	1	20
Penile shots	5	10	8	12	11	3	2	51
Pump	11	10	13	11	6	5	0	56
Urethral suppositories	1	1	0	4	0	3	0	9
Penile implant	2	1	0	0	1	0	2	6
Oral medication (provide medication name in comment box below, if possible)	53	22	24	11	9	3	0	122
Other	2	0	1	0	0	1	1	5
Oral medication name								124
						answered question		180
						skipped question		21

Oral medication was used most frequently at first, but mutual masturbation and oral sex were used more frequently over time. A question remains on why the use of oral medication dropped off so radically.

The data needs to be analyzed by age and treatment modality.

13. How important do you think is penetrative sex is?						
Answer Options	Not important	Not very important	Somewhat important	Important	Very important	Response Count
For you	16	26	56	55	33	186
Your partner	15	10	29	55	69	178
					answered question	189
					skipped question	12

When we examined the data from both the survivors and their partners, we found that the survivors felt that their partners were less bothered by the lack of penetrative sex than they actually were. When the partners were matched, however, the partners were actually less bothered than their survivor partner guessed. There is a conundrum here as one would expect the matched partner results to be closer!

14. Would you want your partner to have a penile implant?

Answer Options	Response Percent	Response Count
Yes	5.9%	11
No	55.3%	104
Maybe	38.8%	73
answered question		188
skipped question		13

Interestingly, 55.3% of the partners were against their partners having a penile implant. More study would be needed to explore the reasons for this."

15. If your partner were to have a penile implant, who is it for?

Answer Options	Response Percent	Response Count
You	8.3%	13
Your partner	33.3%	52
Both of you	58.3%	91
answered question		156
skipped question		45

16. How important is it to experience an orgasm on a regular basis even if it does not result from penetrative sex?

Answer Options	Not important	Not very important	Somewhat important	Important	Very important	Response Count
For you?	14	26	56	55	36	187
How do you think your partner would answer?	17	15	28	64	59	183
Comment						24
answered question						187
skipped question						14

Q16. Respondent Comments

1	12 months after last Lupron shot, orgasms happen only about 25% of the time
2	After ADT ----- no libido
3	desire has also been affected by the lupron for him
4	Early on it was very important with much more emphasis on penetrative sex and orgasm, as his disease progressed it became less important and eventually stopped for both of us altogether.
5	He cannot always have an orgasm now.
6	He has always stated that my pleasure was his pleasure
7	He has basically given up sexually, feels like it just won't work, and it's all just for me now anyway.
8	He has experienced impotence off and on since 1990.
9	He has no lilbido
10	he takes failure personally

11	hsband has completely withdrawn from any form of intimacy
12	I really hesitate to answer for him.
13	It is important to him for me to reach orgasm
14	Larry knew how to arouse a partner quite nicely.
15	my partner has lost his sex drive
16	My sex drive has fallen & husband's has since brain tumor surgery 5 years ago
17	neither has libido
18	Prior to surgery, I had 3 to 7 orgasms per week. Afterwards, the sadness and feelings of loss have interfered somewhat with my ability to receive pleasure from my spouse. But it is improving.
19	SAME
20	Sexual intimacy is more important to me than an orgasm
21	SOMEWHAT FOR ME VERY IMPORTANT FOR HIM
22	using sex toys is fine
23	Very Important
24	We both want to please each other. My husband is just beginning to enjoy it more again.

17. How often do you have penetrative sex now without any form of assistive aids: pills, pump, penile shots, urethral suppository or penile implant?

Answer Options	Response Percent	Response Count
Not at all	76.2%	144
Less than once a week	14.8%	28
1 or 2 times a week	5.3%	10
3 times a week or more	3.7%	7
	answered question	189
	skipped question	12

Out of 93% of respondents who answered this question, 76.2% did not experience penetrative sex. This data needs to be analyzed by treatment and age.

18. How often do you have penetrative sex now using assistive aids: pills, pump, penile shots, urethral suppository or penile implant?

Answer Options	Response Percent	Response Count
Not at all	56.5%	105
Less than once a week	25.8%	48
1 or 2 times a week	14.5%	27
3 times a week or more	3.2%	6
	answered question	186
	skipped question	15

Responses from 92% of partners who answered this question showed a decrease in the number of partners not having sex at all. This data needs to be analyzed by treatment and age.

19. Which of these forms of assistance resulted in an orgasm with or without penetration?

Answer Options	Tried	Did not achieve orgasm	Achieved orgasm infrequently	Achieved orgasm most of the time	Achieved orgasm always	Response Count
Nothing	16	15	5	10	3	46
Self masturbation	18	11	10	29	39	93
Mutual masturbation	13	14	19	42	37	115
Oral sex	11	12	21	34	32	100
Sex toys such as vibrators	10	7	7	15	15	47
Penile shots	8	6	15	11	8	44
Pump	9	9	10	11	6	40
Urethral suppositories	3	3	0	1	1	8
Penile implant	0	1	1	1	3	6
Oral medication	18	9	19	24	14	76
Other	1	1	1	0	0	2
Oral medication name and/or comment						64
					answered question	174
					skipped question	27

The most frequently used method of achieving an orgasm, most of the time or always, was mutual masturbation, followed by self masturbation then oral sex and oral medication. Calculating the success rate in percentages for each method: self masturbation led with 73%, mutual masturbation 69%, oral sex 66% and oral medication 47%. It is interesting that not only was oral medication used less frequently, but it also trailed when it came to achieving successful orgasms.

It would appear that the methods we probably used as adolescents are the most helpful in solving the problem of achieving an orgasm. There are many ways of achieving an orgasm and this is a personal choice. Individuals are most successful if they communicate and work within their personal comfort zones.

This data needs to be analyzed by treatment and age.

20. If you have tried alternative means to reach an orgasm, was it for?

Answer Options	Response Percent	Response Count
You	22.9%	35
Your partner	17.6%	27
Both of you	59.5%	91
	answered question	153
	skipped question	48

It would be interesting to compare these findings with those from the survivors.

21. Can a man have an orgasm without ejaculating?

Answer Options	Response Percent	Response Count
No	1.1%	2
Yes	87.6%	162
Not sure	11.4%	21
	answered question	185
	skipped question	16

Question #21: Orgasms without erections or ejaculations are possible.

22. Can a man have an orgasm without an erection?

Answer Options	Response Percent	Response Count
No	10.7%	20
Yes	55.6%	104
Not sure	33.7%	63
	answered question	187
	skipped question	14

Comparing these answers to those of the survivors would be interesting. It is of interest that 44.4% of the partners did not know, or were unsure that orgasm without an erection is possible.

23. Would you agree that women get a great deal of security and sexual pleasure from pure intimacy, kissing, hugging, holding hands, lying close together?

Answer Options	Response Percent	Response Count
No	3.7%	7
Yes	92.1%	176
Not sure	4.2%	8
	answered question	191
	skipped question	10

Both the survivors and their partners endorsed this finding.

24. What percentage of women do you think experience vaginal orgasm?

Answer Options	Response Percent	Response Count
0%-10%	7.1%	13
11%-20%	16.5%	30
21%-30%	23.1%	42
31%-50%	22.0%	40
51%-70%	20.9%	38
71%-90%	9.3%	17
91%-100%	1.1%	2
	answered question	182
	skipped question	19

Experts state that only 24% of women experience vaginal orgasms.

25. Did you both go to the appointments with your urologist?

Answer Options	Response Percent	Response Count
No	12.2%	23
Yes	64.6%	122
Occasionally	23.3%	44
	answered question	189
	skipped question	12

Two heads and two sets of ears are a good idea. The word "cancer" frequently stops the thought process.

26. Did your treating physician discuss any of the following?

Answer Options	Yes	No	Not sure	Response Count
That after treatment male sexual function would be different requiring some adjustments.	101	62	17	180
That women need to be wanted and needed physically	9	148	19	176
That for women foreplay provides closeness and intimacy.	11	148	16	175
That orgasms are important for women.	12	149	14	175
That few women have vaginal orgasms.	4	155	12	171
That clitoral orgasms are stronger than vaginal orgasms.	2	158	15	175
That manliness and masculinity are closely tied to having an erection and that this issue is manageable.	28	130	17	175
Libido remains the same even if sexual function does not or partially returns.	18	131	26	175
Male orgasms will be dry. (no ejaculation)	91	68	17	176
That possibly penetrative sex may not end in an orgasm.	17	129	26	172
That is is important that partners maintain intimacy, hugging, kissing, caressing, lying naked together.	24	132	19	175
How often you had sex.	30	125	19	174
My physician discussed erectile dysfunction issues.	100	62	15	177
			answered question	181
			skipped question	20

There are a lot of "No" responses here. It has been suggested that it is a good idea to go in to see the treating physician armed with the questions you need answered. Prostate cancer websites could offer a list of possible questions.

27. If your physician discussed with you the possibilities of having problems with sexual function after treatment, was there a percentage of problem possibility provided? MUST BE WHOLE NUMBER NO DECIMALS!

Answer Options	Response Average	Response Total	Response Count
Percentage	50.00	3,050	61
		answered question	61
		skipped question	140

Responses ranged from under 10% to 100%. So what happened to expert information? Only 14% of survivors answered this question. It would be interesting to explore possible explanations for this.

28. Overall how big a problem do you consider your sexual dysfunction to be for you?

Answer Options	Response Percent	Response Count
No problem	15.0%	27
Very small problem	16.1%	29
Small problem	18.9%	34
Moderate problem	31.1%	56
Big problem	18.9%	34
	answered question	180
	skipped question	21

This data needs to be analyzed by treatment and age.

29. If your partner is on androgen deprivation (in the form of shots or similar treatments to depress or eliminate your testosterone) are you able to have an orgasm with a:

Answer Options	Response Percent	Response Count
Normal erection	1.9%	3
Assisted erection	8.9%	14
No erection	15.3%	24
Does not apply	73.9%	116
	answered question	157
	skipped question	44

This data requires some innovative interpretation. Only 35% of the survivors answered this question, and 116 were not on androgen deprivation medication (see "Does not apply"). For the other 41 (3+14+24) partners who answered the question, we can assume that their survivor partners were on androgen deprivation medication. Only 1.9% of the men had a normal erection. Did the 44 partners who did not answer this question

143

have survivor partners who were not on androgen deprivation medication? This data hasn't been broken down by treatment or age.

There is some excellent research on androgen deprivation taking place in both Canada and United States. Information is being developed on the best methods of providing help for dealing with the side effects of this form of treatment.

30. How big a problem do you think the sexual dysfunction is for your partner?

Answer Options	Response Percent	Response Count
No problem	8.6%	16
Very small problem	10.2%	19
Small problem	16.1%	30
Moderate problem	32.3%	60
Big problem	32.8%	61
	answered question	186
	skipped question	15

This data needs to be compared to the survivors' answers to this question. In addition, the answers from the matched couples need to be evaluated.

31. Where did you obtain your information on prostate cancer?

Answer Options	Response Percent	Response Count
Library	11.7%	22
Magazine	9.6%	18
Book	52.1%	98
Internet	75.0%	141
Cancer support group	37.8%	71
Friend	9.6%	18
Prostate cancer survivor	30.3%	57
Physician	54.3%	102
Other	10.1%	19
Other (please specify)		37
	answered question	188
	skipped question	13

Most partners obtained their information from the Internet. The next most common sources were physicians, books, cancer support groups and other prostate cancer survivors.

32. Support Groups If the answer to "a." is "No" go to "b." If the answer to "b." is "No" go to "c." If the answer to "c." is "No" go to the next question.

Answer Options	No	Infrequently	Often	Very often	Response Count
a. Did you attend a support group prior to treatment?	154	8	8	7	177
Did your partner go with you to this group?	52	8	7	8	75
Did this group help you?	40	4	6	11	61
b. Did you attend a support group after treatment?	106	22	16	19	163
Does your partner go with you to this group?	40	9	15	26	90
Did this group help you?	36	10	19	21	86
c. Do you attend a support group now?	130	13	11	13	167
Does your partner go with you to this group?	49	10	9	15	83
Did this group help you?	35	8	16	15	74
				answered question	179
				skipped question	22

88% of partners answered this question. Examining this data, it would appear that support groups were not the initial source of help and/or information. Of those who answered the question, 19% attended support groups after treatment; 45% of these partners felt they were helpful, and 45% of the survivors also found them helpful.

33. If you attended a support group, kindly answer the following?

Answer Options	Response Percent	Response Count
Could you describe how the support group helped or if it didn't, why it did not help?	93.2%	55
Could you add any suggestions for improvement of the support group?	71.2%	42
	answered question	59
	skipped question	142

Q33 part 1: "Could you describe how the support group helped or if it didn't, why it did not help?" Respondent Comments

1	? 32 also doesn't really give answers to help, we just began attending a group and there isn't a yes answer to fill in

2	1st cancer support group was GREAT, however we moved and 5 years later prostate cancer returns and now held only for the survivor in the form of meeting once a week for 5 weeks reading topics in a book and discussion with a GREAT Nurse leader Picking Up the Pieces - Moving Forward
3	Ability to discuss specific issues, current treatments & feelings about the disease and side effects from treatments. The women would often break away from the men for the last 45 mins to discuss personal feelings, occasionally we had our own meetings without the men to discuss partners issues in coping with prostate cancer.
4	all older men - topic of discussion was HEART health - was looking for COUPLES guidance/knowledge specific to intimacy.
5	Although I haven't attended I enjoy reading the "Us Too" Newsletter and hearing about the discussion from my husband.
6	being there made me angry because it confirmed my feeling that my husband received delayed inadequate treatment
7	Caring ,interest, friendship, information,
8	Dr speaking
9	had a couple of internet support groups
10	helped to find highly skilled surgeon in our area
11	I attend a support group. It is very helpful as we are in limbo about my husband's health care treatment.
12	I belong to Us Too on line, not any "physical" support group
13	I feel the problem was no knowledge of support groups at time of diagnosis. It would have been nice to start then.
14	I know there are others who have similar concerns. I learn more about my husband's disease. I/we have support.
15	I only went a couple of times with my husband. The literature helped the most. It helped my husband tremendously.
16	informal contact with doctors on the staff of the treatment center
17	information sharing, lending library, questions to ask doctors, things to think about/consider
18	Information...literature...books....speakers....group discussions
19	Information; library (videos/books/pamphlets); guest speakeers; womens group after main meeting very helpful
20	Initially helped, gave a lot of info, then had men who steered every discussion to their issues, so my husband did not feel free to discuss what he needed help with
21	Intimacy was not discussed openly and there was a huge age difference
22	It talks about vitamins and stuff like that
23	It was good to be able to hear what other men were experiencing & exchange information
24	it was good to see other men surviving the cancer and living happy.. We were the youngest members of the group
25	It was informative for questions regarding treatments, symptoms, etc. Never attended a meeting where sexual questions were discussed.

26	It was very informative and open. We were able to obtain information, as well as ask questions.
27	Just the support from the group helps, plus their sharing their own experiences help
28	Just to know we were not alone helped a lot.
29	Learning about experiments and new methods of treatment on the horizon
30	Most attendees were surgical patients - those of us dealing with ADT are in the minority and have a bunch of very different issues
31	My husband is quite young (54) and the other men there were in their 70's. My husband went to one meeting but refused to go to any more because he didn't want to hear what was going to happen.
32	My ladies quilting group can talk about and help solve most problems.
33	No one has suffered the mutilation and abuse my husband has - nothing in common with support groups.
34	online support group of women whose partners had prostate cancer
35	Open and frank discussions, confidentiality, humor, fellowship and sharing information
36	opened up communication between us
37	other guys shared their experiences
38	people kind but much older than me, so didn't really understand my personal problems
39	provided us with very informative information to help with our forms of treatment that may be required
40	Share info and listen to others and increase knowledge also speakers were interesting and informative. Comforting
41	source off information and reassurance
42	Subject matter at time was not relevant
43	The group was mostly a pep rally, with no discussion of specific problems after treatment.
44	The support Group was strictly controlled and didn't allow for too many questions.
45	The support group was too Christianity based for my taste
46	The support group we attend is for all types of cancer
47	Things were discussed in a mixed group and I was very uncomfortable and didn't want to return.
48	Very open--talk about many aspects. Help one another and new members.
49	We attended the XXXXX group, but it was geared toward the older gentlemen who were not sexually active.
50	We didn't feel so alone in our problem. We realized that other people were having the same issues that we were having.
51	We got a feeling of understanding and support
52	We were in the process of deciding whether surgery or radiotherapy would be the best form of treatment and the discussion and frank answers to our questions were helpful.
53	We were the youngest couple there and it did not appeal to us

54	Women were honest about their fears and problems. There was a feeling of not being alone.
55	You see others with same problems. Get good information.

Although only 29% of partners answered this question, 93% commented on the benefits of support groups. Overall, they felt the information was informative and helpful. It comforted them and decreased their feeling of being alone. They reported that often there was a "huge" age gap; liked separate partner groups; and felt pre-treatment groups would be helpful. Online support groups were helpful.

Q33 part 2: "Could you add any suggestions for improvement of the support group?" Respondent Comments

1	A more flexible meeting with more chance for involvement.
2	A support group for spouses and partners would be welcome.
3	address the reality that some women will never have sex again
4	An 'orientation' guide with useful links to the DIFFICULT questions - a list of clinical sex therapists specializing in prostate cancer-related challenges.
5	Ask women to share their experience dealing with PC
6	Be supportive and provide info also the importance of nutritrion should be stressed.
7	Couples should start attending before treatment for prostate cancer.
8	Educate doctors to spread the word or have literature in their offices about support groups.
9	enjoy the women's break out meeting as it to provides information for the women
10	Gear it toward the younger generation who are still sexually active.
11	Head of it needs to have more control over where discussion heads, and allowing all present a chance to talk without interruption
12	I am surprised why I am usually the only wife there.
13	I don't think women should attend the meetings except if the subject includes them It is hard enough to get men to open up but with other women there it is not going to happen.
14	I longed for support from the cancer community, but it never came.
15	I love my support group.
16	I would like to attend one but our local group does not discuss intamcy issues-they are older- the Toronto groups would be good but we live too far away
17	Include discussions about incontinence, ED, depression, marriage difficulties that can result from those
18	It would be best to have a doctor who is able to volunteer some time to attend meetings....someone who can tell it like it really is...
19	Less religion, more actual information and support from each other and facilitators

20	Men just won't talk about feelings. Maybe wives (or partners) would. But not in a small town.
21	More information for couples regarding meeting the sexual and emotional needs of each other prio to treatment and afterwards. Also, something from a female prospective to advise husbands and significant others. Maybe a female urologist.
22	More intimate discussion about erectile dysfunction or a separate group to discuss this
23	More wives to attend and enter into the conversation
24	more women my own age-40's or 50's
25	Need more speakers and sometimes the groups where just to big and it was difficult to hear.
26	No, Our support group is wonderful and very open.
27	No. It works for me.
28	not really
29	one for partners regarding lack of sex.
30	Only for me to have time to attend more often.
31	only went once because i was so upset
32	Our group of wives disbanded for lack of leadership. Women need support groups and there isn't one now.
33	ours was canceled due to funding issues
34	Perhaps pairing him up with one person that he could talk to instead of a group situation would have helped him to open up and ask for help and information - instead of having to ask in front of several men AND their wives.
35	Possibly separate or additional support sessions for ADT patients and spouses
36	The meetings where the men & women were divided into seperate groups enabled us to get down to the real nitty gtritty stuff
37	the presenters were medical staff and the group talked about their experiences with their mates
38	there needs to be more openess about sexual problems with treatments
39	They really didn't offer anything for spouses
40	This support group has sessions for prostate cancer survivors but I do not accompany my husband for those meetings
41	Yep, I am left out of it - guess only SURVIVORS have issues - not a separate meeting with someone at the same time in another room addressing the FEMALE issues - I am a NOBODY...
42	Yes, greater outreach to men who are newly diagnosed

Of the 29% of partners who answered the question on support groups, 71% made suggestions on how to improve the groups. They felt sexual issues should be discussed and even a list of sex therapists should be made available. They felt that groups for younger people should be made available, and that too few partners were encouraged to attend meetings. Meeting information should be made readily available and somehow the newly diagnosed should be encouraged to attend the meetings. Women felt left out.

34. Do you make use of the internet as a support network?

Answer Options	Response Percent	Response Count
No	43.8%	81
Yes	56.2%	104
Comments		47
	answered question	185
	skipped question	16

A smaller group of partners than survivors used the Internet to search for information.

Q34. Respondent Comments

1	After this whole prostate problem was over, our "Dr" didn't even call us with the results of the pathology report.... do you know why? Because he messed up! The prostate had ZERO, yes zero % cancer. My husband has Acute and Chronic Prostatitis. This is why it is SO difficult to deal with. My husband could have been treated with a month supply of antibiotics... life will never be the same, and our insurance company doesn't cover the very expensive medication. (Another issue that was not discussed by the "Dr'".
2	again my husband did this not me
3	Am looking for support network now.
4	Because my story is published on the internet on the da Vinci website, I often get emails with questions which always respond
5	blogs and many websites
6	But my partner does. I have used it for information.
7	Did for the first year, getting support from other survivors and partners, but have no need for it now. Others' frank comments and advice was very helpful.
8	Did not realize your site until a friend from the 2005 cancer alerted us to this site.
9	Found a wonderful forum for wives and partners of PCa patients.
10	I correspond with a very helpful woman and read postings on a couple of websites.
11	I have found this virtual support network extremely valuable for candid information and details that the doctor overlooked.
12	I have just joined an online forum for ladies only
13	I like P2P with Stephen Strum--I think he gives excellent advice and almost all men could find cases similar to their specific one there.
14	I looked things up on the internet. I tried to find a support group but was unsuccessful; all I really wanted was to talk to other women using caverject with their partners. I did feel pretty alone about it, although it worked for us.

15	I read Dr. Strum's P2P letters. They are just about treatment though. My husband won't discuss anything with me.
16	I read what others have to say.
17	I seldom use the computer
18	I think my husband does, for sure he did before and during treatment
19	I try to find real talk from women partners but haven't found any yet.
20	I use facebook and look up information from other survivors..at different websites and ordered bracelets
21	I visit ladies prostate forum and discuss issues with the other women in similar situations. They are a valuable source of support and information
22	I was able to connect with other women who had formed an Internet Support group of sorts. THis was helpful.
23	It's been 11 years since the surgery and we still are helped by the internet groups (PCAI & RP). We wish we'd found them before the surgery because we'd have been more prepared for difficulties & ways to minimize them.
24	just started
25	Just started this week.
26	Lifesaver-lots of good information. New information too.
27	Many prostate survivor websites
28	my husband does
29	my husband has found them very helpful
30	My husband has used the internet considerably.
31	no time to pursue
32	Not recently, but immediately after the surgery.
33	Particularly since there is little to no support information from physicians.
34	Run a prostate cancer motorcycle charity fundraiser
35	Searching for personal experiences by wives that are similar to mine that I can identify with.
36	Seedpods
37	The new prostate cancer info site, Dr. McHugh's blog, Healing Well forums.
38	The women on the group are not helpful because they want they to remain positive and primarily interested in the physical and emotional effects of the cancer on their men. We are a support for them.
39	There is a women's only forum which has very useful information and is a good place to see what other women are feeling and experiencing.
40	There's some good information out there.But it can be overwhelming if you overdo it, especially for the person who's got cancer.
41	they were helpful
42	this is a small part of my life. i live for the future not the past. you just need your imagination to have a great sex life. people get stuck on what was not what is.
43	VERY HELPFUL
44	We are members of also on the board of directors of a very active support group

45	we make use of the internet for finding information
46	Without internet support I would have nothing. The online groups and information have been invaluable resources.
47	Yes it has helped but when we were diagnosed Xxxxxx already had extensive bone mets and lymph nod involvement. So a done deal. Angry he was not diagnosed sooner so he could be cured - let down big time!!!!!!

35. Have you found internet sources helpful?

Answer Options	Response Percent	Response Count
Not helpful	8.4%	14
Not very helpful	13.3%	22
Helpful	43.4%	72
Very helpful	34.9%	58
answered question		166
skipped question		35

Of the 166 of partners who responded, 78.3% found the Internet helpful or very helpful. The education level of the partner respondents was high, which could be considered a bias.

36. What do you see as the weakness of internet research?

Answer Options	Response Count
	115
answered question	115
skipped question	86

Q36. Respondent Comments

1	you're only reaching people who use the internet and are open to this kind of communication ie generally more liberally minded individuals
2	You need to do extra research to make sure the info is correct.
3	You have to be VERY careful that the information is from a scientific/professional source. Commerical attachments are annoying.
4	You have to be careful of the sources
5	Wondering if the info is correct. Sometimes it's not always current.
6	WHO TO TRUST
7	very clinical/ not personal enough, find it hard to relate....mostly american...not much australian content
8	Validity of information.
9	Uncertainly regarding reliability of information

10	trusting the source to be accurate and up to date
11	Too much to focus
12	Too much out there; hard to sort through the clutter
13	Too much information, most of the times very negative.
14	too much information for those that don't know how to disseminate nor understand the information
15	Too much information can overwhelm a person.
16	Too much information at times.
17	too much information
18	Too much information
19	Too many sources without supporting documentation
20	too many sites and different answers
21	To much bad information
22	Time, vision and not knowing what to look for.
23	time consuming
24	They skip over the likihood of impotence -- too lightly.
25	the possibility of poor information and the likelihood of slanted information provided by for-profit sources
26	The need to be alert for inaccurate information.
27	subjective commentary
28	sometimes too much information to process
29	Some sites 'gloss over' the issues - some people want direct address, ex: What can I do to make my partner want to have sex again?
30	Some sites are difficult to negotiate
31	Some internet sites obviously promote a certain kind of treatment. People suffering Prostate Cancer tend to look for things that sound good in preserving urinary/sexual function....freezing surgery/....high frequency sound being a couple. These things are more or less in the preliminary stages. My conclusion was surgery was best, the more I read and searched.
32	Some information is not documented, and alot of emotional stuff is out there
33	Some are not reliable...accurate...realistic
34	so much information and some of it is not good information
35	sifting through shear mass of possible sources and the reliability of said sources
36	One has to understand that there is alot of "wrong" info out there - you can't believe everything you read.
37	One has to discriminate between good and inaccurate information.
38	One has to carefully evaluate the sources.
39	Nothing
40	Not the same as being in person
41	Not sure it can be trusted.
42	Not sure
43	not sure
44	NOT SURE
45	Not specific to our situation.

46	not specific
47	Not really tailored to our situation-we get every possible angle on the problem and cannot really tell if it applies to us or not.
48	Not personal
49	not necessarily reliable; information may not be true
50	not enough feedback on problems as a whole
51	Not enough detail or too much related information so don't find an exact search
52	Not easy to find specific information since treatment and response varies
53	Not being able to guage a person's reactions.
54	not always true
55	Not always specific to my situation.
56	Not always easy to find good info
57	None
58	None
59	no real person to converse with
60	No personal contact and understanding
61	No live person to talk to whose husband went through this.
62	No
63	Need more discussion forums - so people can share experiences, and success stories with others to give hope
64	N/A
65	My husband does not read or research as well
66	Misinformation! Also, some groups seem to exist for the sole purpose of whining then trying to share helpful hints/information which is quite depressing.
67	Maaybe too much info which can confuse the decision making
68	lots of misinformation--need to be selective
69	Lack of the facts from a medical professional
70	lack of relevant information
71	Lack of coverage of the emotive issues around significant changes to sexual activity
72	Lack of consistency or centrality of information. Lack of clarity regarding penile rehabilitation protocols and treatment protocols.
73	Lack of commpassion for spouses of prostatectomy patients and the isolation.
74	Lack of 3-D interactions
75	Knowing what site is valid. I trust US Too completely.
76	Know that you are using legitimate sources
77	It's limited
78	it was very sad to read the painful stories
79	It is difficult to be able to share your whole situation and experiences when answering very specific questions.
80	It cannot go into specific or personal questions
81	Issues not relevant
82	Integrety of information

83	information is out of date, medical info has to be paid for
84	Inconsistency of factual evidence
85	Impersonal
86	If you don't have access to medical journals it limits you. Sometimes too many choices
87	I haven't found much about sexuality after radical surgery where the nerves couldn't be spaired.
88	I have not had time to research. No computer at home
89	I don't use it
90	I don't know what sources I can really trust, the same way I feel about it in general. You have to do lots of research to begin to get the whole picture.
91	I don't ever see any.
92	I don't believe everything that is offered on it is necessarily true.
93	I do not like to share personal information
94	I am not internet saavy
95	how factual is it?
96	Hard to search
97	Hard to gauge validity of information, often too much of it.
98	hard to access medical information if not a clinician
99	getting access to accurate information.
100	General information / doesn't take into account pre-existing challenges.
101	face to face discussions provide more reality......more needed affirmation
102	Each country has different medical laws and information.
103	don't weigh in the emotional factor
104	Does not provide info to ask a urologist and there are not pamphlets that urologists give out with options
105	Doctors seem to discount what Dr. Strum says: "Status begets treatment"
106	Conflicting information
107	cant really say I see any weakness
108	Can't always tell if the information you get is correct.
109	can not rely on accuracy
110	authenticity of research, reliability of info sources.
111	Authenticity
112	Any information should be verified by other means...physician, etc.
113	all internet searches around erectile dysfunction seem to be geared towards regaining erection
114	Accuracy of data
115	accuracy can be iffy

Of the 57% of partners who responded to this question about the weaknesses of the Internet, 100% shared their opinions. This is an impressive first, and we thank you.

Of those partners who commented, 27% felt that the trust factor of the information was lacking as was the reliability, validity and current content; 16.5% felt that the data was biased, negative, incorrect and unsupported and needed to be verified, which was time-consuming; 14% felt that there was far too much data and some of it was bad and unsupported; 11% felt it was impersonal and not the same as face-to-face discussions. The issues of impotence and how to deal with this were not addressed. It was hard or costly to access medical papers; there were also too many "pop ups" and for-profit advertisements.

37. What do you see as the strength of internet research?

Answer Options	Response Count
	114
answered question	114
skipped question	87

Q37. Respondent Comments

1	Yes
2	Wide variety of resourses.
3	We became educated on the problems facing us, and could understand the potential situations which might confront us.
4	Very accessible
5	variety of viewpoints
6	Use groups/medical sites that have credible, knowledge information. One can compare info & use that to make wise decisions. Also, helpful knowing what you're facing is not unusual & you can learn from someone elses experience.
7	uptodate medical advice
8	Up-to-date coverage of the medical advances for the treatment of Prostate Cancer
9	Unlimited amount of information.
10	too much information
11	There is lots of information out there.
12	There is a wealth of information, including some of the latest research.
13	The wealth of information out there and the ability to quietly read up and absorb and understand
14	the same
15	The information available gives a starting point for researching the subject.

16	The availability of support & sometimes advice very quickly, usually within 24 hours.
17	That it's the only sourse of information and comfort available.
18	Support
19	share info
20	Seeing that other are going through similiar problems.
21	Resources, people experience comments
22	REasy access and privacy.
23	real people testimonials
24	quick, can be done at one's own time availability, variety of input
25	Provides at least some info
26	private - easily tackles the issues that are 'delicate' to discuss face to face.
27	Private
28	Privacy
29	Perhaps more freedom to talk - anonimity (sp?)
30	Opportunity for many, many people to respond with anonymity.
31	One can gather a tremendous amount of information - then follow it up.
32	not sure
33	NOT SURE
34	None
35	Newest info available, how others are dealing with info, great suggestions for moving forward
36	Multiple websites to address various needs and issues.
37	Multiple sites for many types of cancer
38	Multiple opinions Always available
39	more information and answered questions often not asked at office visits
40	More and more high quality information is available in all fields on the internet.
41	many resources
42	magnitude of information
43	Lots of sources of information easily available.
44	Lots of information.
45	Lots of information which would otherwise be hard to find
46	Lots of information out there
47	lots of different voices-opinions
48	Lists the many treatment options; but hard to tell which works best if you are young and still sexually active
49	knowledge is strength - its better to know what you are up against
50	Knowledge
51	Keeps our awareness up and we know what questions to ask our doctor
52	Its immediacy
53	it's fast and easy and updated regularly
54	It's easy

55	It very likely provides some options to try when traditional methods are not effective. Not all the stats are the same from study to study so you need to take a balanced apporach to reading this literature.
56	It reaches a lot of people.
57	it is very informative. A lot of my questions were answered via the internet--especially with people being truthful.
58	It is always available when you have a question.
59	It gives an average idea
60	It can be done in the privacy of one's home.
61	Information available when you want it
62	Information and what to expect
63	Information abounds and there are numerable links.
64	immediately accessible
65	immediate answers to questions
66	I know that I'm not the only one going through this.
67	I don't use it
68	Helps you help yourself & talk better with your doctors
69	Helps to know others have similar problems
70	good explanations
71	Gives us information to talk with our doctor and others
72	get to see alternatives, guidance, encouragement
73	finding the right webiste with accurate infomation
74	find out answers needed by real people who actually are going through the same things
75	Everything is available, right in front of you
76	enormous amount of information easily retrievable
77	Easy-Lots of info
78	Easy to get info.
79	Easy to access
80	Easy access and to answer specific questions.
81	Easily obtained
82	ease of access
83	don;t know yet
84	Discretion.
85	consistent involvement and support from numerous survivors
86	comments from patients, their reactions to various drugs and treatments.
87	CAN FIND OUT ANYTHING YOU LIKE
88	can do at home
89	Can be on cutting edge of new developments.
90	But very good neutral source, serves as a base for educated discussions with the urologist
91	Being able to read at my leisure.
92	being able to discuss private things without being embarrassed
93	available for something
94	Available 24/7
95	availability of current research

96	*Anytime day or night*
97	*Anonymous*
98	*Anonymity of responders makes for more openness.*
99	*Anonymity not having to face another with personal info.*
100	*Anonymity*
101	*Accsessable quickly and to a wide range of information from various sources*
102	*Accessibility to personal stories of people who have actually been through the diagnosis, treatment and recovery.*
103	*Accessibility*
104	*accessability, broad and wide information*
105	*access to most current information*
106	*access to medical websites*
107	*Access at anytime*
108	*able to reach a lot of people from a wide demographic range*
109	*able to be open with some anonimity*
110	*ability to connect one to one*
111	*Ability to access specific information on treatments, trials, drug information and side effects, and support group info.*
112	*A lot of information easily available*
113	*A large myriad of information*
114	*24/7*
115	*Yes*

Although only 56% of the partners answered this question, 100% of them responded about the strengths of the Internet and shared their opinions. This is impressive and, once more, we thank you.

Of those who commented on the benefits of the Internet: 19% felt the wide variety and amount of available data was positive; 15% cited its speed and accessibility; 12% mentioned privacy to do "self research"; 12% felt the number and speed of responses were invaluable; 9% felt that the information and educational aspects of it were helpful. Other comments included that the Internet was a resource 24/7, and that information was current and up-to-date and often laid the groundwork for developing questions to ask the treating physician.

38. Do you have an informal support network?

Answer Options	Response Percent	Response Count
Wife	10.0%	11
Friend	60.0%	66
Prostate cancer survivor	29.1%	32
Other	32.7%	36
Other (please specify)		59
	answered question	110
	skipped question	91

Among the partners, friends rate highly as support.

39. How beneficial has the informal support group been?

Answer Options	Response Percent	Response Count
A hindrance	1.8%	2
Not helpful	9.1%	10
Somewhat helpful	31.8%	35
Slightly helpful	13.6%	15
Very Helpful	43.6%	48
If you checked a hindrance please explain		5
	answered question	110
	skipped question	91

Q39. Respondent Comments to "If you checked a hindrance please explain"

1	no help
2	No support
3	Sick of people saying it is the best cancer to get, or that it is easily treatable. Even in the family, they don't really understand the reality of the post-op side effects
4	There isn't any for the females
5	We as a couple have been very helpful to each other.

Of 54% of the partners who answered this question, only 4.5% made comments. In general, they felt their informal support was not helpful, except for a couple who provided support for one another.

40. How do you prefer to get prostate information?						
Answer Options	Never	Seldom	Occasionally	Often	Always	Response Count
Email	32	11	32	32	15	122
Blog	71	12	17	9	4	113
CD	76	10	9	2	1	98
DVD	70	9	11	8	4	102
Web Video	62	15	15	3	4	99
Magazines	34	22	36	16	6	114
Newspapers	32	26	32	14	2	106
Library	45	18	29	11	8	111
Internet	11	7	41	61	31	151
Support Group	44	13	19	22	16	114
Facebook	86	8	4	2	1	101
Twitter	96	1	0	1	0	98
Doctor	13	15	40	39	28	135
Prostate Cancer	29	18	29	27	22	125
TV	45	26	24	4	3	102
					answered question	173
					skipped question	28

Of the 85% of partners who answered this question, 53% preferred getting their information from the Internet; 39% from their physician; 27% from other survivors; 27% through email and 21% from their support group. Newspapers, magazines and libraries appear to be secondary sources of information.

Support groups need to increase their efforts to provide superior support to prostate cancer survivors and their partners.

41. When there were erections with sexual stimulation, how often were your partner's erections hard enough for penetration? WITH NO MEDICAL AIDS OR ASSISTANCE!

Answer Options	Response Percent	Response Count
No sexual activity	22.7%	41
Almost never or never	35.4%	64
A few times (much less than half the time)	14.4%	26
Sometimes (about half the time)	5.0%	9
Most times (much more than half the time)	9.9%	18
Almost always or always	12.7%	23
answered question		181
skipped question		20

This data needs to be analyzed by treatment and age. An important question is how it compares to what the survivors report.

42. When sexual intercourse was attempted, how often was it satisfactory? WITH NO MEDICAL AIDS OR ASSISTANCE!

Answer Options	Did not attempt intercourse	Almost never or never	A few times (much less than half the time)	Sometimes (about half the time)	Most times (much more than half the time)	Almost always or always	Response Count
For you?	38	59	21	17	19	26	180
For your partner?	32	51	25	11	27	24	170
answered question							182
skipped question							19

This data did not come from matched partners but it would be of value to compare the survivors' answers to these responses. The data needs to be analyzed by treatment and age.

43. When there were erections with sexual stimulation, how often were your partners erections hard enough for penetration? WITH MEDICAL AIDS OR ASSISTANCE!

Answer Options	Response Percent	Response Count
No sexual activity	17.0%	26
Almost never or never	19.0%	29
A few times (much less than half the time)	11.1%	17
Sometimes (about half the time)	9.2%	14
Most times (much more than half the time)	18.3%	28
Almost always or always	25.5%	39
answered question		153
skipped question		48

This data needs to be analyzed by treatment and age. It would also be helpful to compare it to the survivors' reports.

44. When sexual intercourse was attempted, how often was it satisfactory? WITH MEDICAL AIDS OR ASSISTANCE!

Answer Options	Did not attempt intercourse	Almost never or never	A few times (much less than half the time)	Sometimes (about half the time)	Most times (much more than half the time)	Almost always or always	Response Count
For you?	23	26	20	12	30	38	149
For your partner?	22	19	17	11	33	39	141
						answered question	149
						skipped question	52

This data needs to be analyzed by treatment and age, as well as in the matched partner group to see how it compares to what the survivors report.

45. General comments/feedback about this survey

Answer Options	Response Count
	71
answered question	71
skipped question	130

Q45. Respondent Comments

1	*A little confusing - but ok. For 10 years - there has been no sex because my husband is androgen independant - with a shot of Lupron every 4 months. Now - he's too old - going to be 79 - even if he were to go off the shots - there is probably no hope for sex. But I really think there was a mistake made somewhere - as he was told by the 1st urologist after his surgery - that he probably had 1 1/2 years to live. His psa - is 0.1 and has never gone any higher. He's quite alive and feels good - no metatases anywhere. We are enjoying life - but no one can tell me that he has needed to be on these injections for 10 years. Gleason score=9. (Don't believe that either)*
2	*After having prostate cancer surgery and then radiation for recurrance, I realized that I would always be impotent. Therefore my wife and I decided that I would have an Inflatable penile implant. I have been 100% satisfied with it and would highly recommend it. We did OK with the pump, less so with injections, and did not like the interuption in foreplay they caused. The implant is not the same as before treatment, but it is close.*

3	Any research outcomes which can be used to help couples dealing with this hateful disease will be welcomed in the spouse/partner community. While there is much info on diagnostics and treatments and psa levels, there is very little detailed information on how to help your partner deal with the heartbreaking emotional trauma of a PCa diagnosis when he is a relatively young (50) and a very sexually active man. Being patient is not going to be enough at some point, and we both know it. It seems superficial to some folks to dwell upon your sex life as opposed to life itself, but that is the way he feels. My man is more afraid of the indignity of incontinence and impotence than he is of dying at this early point. It scares me more than I can say. Thank you for this important work you are doing for all of us.
4	Assume medical aids includes medication such Viagra. It would be nice if there was a support group for wives. We have very few wives who attend our group meetings and almost all never say anything.
5	At my age I have been living without penetrative intercourse for ten years, since my husband's radical prostatectomy. I prefer that to cancer death had he not been operated.
6	Being on hormone therapy, my husband has no libido. He is rarely interested in sexual activity. He is still able to get a partial erection but no orgasm.
7	Don't know the answer for my husband
8	First the spontaneity is gone...he has NO desire, I have had to ask for sex. He was afraid to say no, because I had been patient for SOOO long. Then I had to wait for the injection or pill "to work", which was like waiting for bread dough to rise. So he went thru the motions very clumsily and uninterested...no foreplay, just a feeble attempt to penetrate...that made me feel worse. I stopped asking because NO sex is easier to accept than disinterested, obligatory fumbling. At first I tried to help him feel 'manly' by doing the things that gave him pleasure..manual stimulation, oral sex,..it took him a while to relax and ENJOY, rather than him trying to perform. But then I was left totally unfulfilled, and feeling some resentment....It's disheartening to think that we will never have SPONTANEOUS, pleasurable,enthusiastic, erotic, fulfilling intimacy ever again. I can't remember the last time he has even slept in the same bed with me, let alone intiated any physical interaction. About a year now... I feel like a part of me has died.

9	Good survey, thank you for targeting these important and overlooked points about impact of the disease and its treatments on intimacy. The most common response from friends or family has been, "At least your husband got a cancer that is curable." There is no understanding of the impact on the quality of life, expression of relationship and experience of grief and loss when it comes to the issues of sexuality and intimacy. Of course I am grateful my husband is alive! But I still have to acknowledge my anger and grief at the impact of the disease and its treatments. Also, although our doctors were skilled surgeons, much improvement could have been made to prepare us mentally and emotionally for what was to come after surgery. I think having a visit with an LPC, LCSW or sex therapist prior to surgery should be a mandatory part of preparation for surgery or other treatments, along with bone scans, ct scans and blood work.
10	Had penile implant a year after surgery because we found shrinkage even using shots as often as recommended. Discussion with dr. doing the surgery we were told that things would improve but the shrinkage was perment. And it was. So intercourse is not satisfactory. For me, the wife. We were 49 years of age when the surgery was performed. So for this survey and others that I have read, I find they lack some important questions. This is one of the better survey's I have and the first I attempted to answer.
11	He doesn't seem to have the desire for sexually activity and it is hard for me to want any hugging, kissing, etc. Each time we tried to have intercourse, and failed, it seemed to add to his feeling of failure. I haven't been able to help in this area. Maybe because I have dealt with his impotence for so long that I have given up.
12	I am a breast cancer survivor with bladder prolapse married to my 10 year survivor of prostate cancer. It took us 8 years to get it "right", despite all our roadblocks. We now have the best sexual relationship in 44 years of marriage.
13	i answered these last questions based on when we had intercourse before my ovarian cancer.
14	I didn't complete this questionnaire because I found the wording very poor- many of the questions seemed to be aimed at the man with the prostate cancer not the partner of the man. I didn't want to give misleading data for your survey.
15	I felt that this survey was GREAT - extremely well thought out. I am so disappointed how the female partner is left out - sorta to 'dry' Non cancer therapists have NO clue what the females are going through and what the man feels like on lupron shots and radiation treatments and how mentally and physically affected. He is also taking: Casodex 50 mg oral tablet one time a day - this 'boosts' the effect of the Lupron for a better track record. He also cannot drink while on this medication - which he has been faithfully doing. I think it has been definitely helpful to me reading the book - wish I has started before and during his treatment of radiation rather than here is the book (when eveything was done)

16	I find the main issue with sexual satisfaction for the partner to be the effects of HDT. My husband has no interest in sex and I am not interested in being "serviced" so unless there is a solution to the lack of libido, I think after three years of this misery our marriage is over. I am intensely angry over the medical community's lack of attention to this issue and more angry that the pharmaceutical approach to PCa is to create a third gender of enuchs with the intention to keep their customers alive as long as possible.
17	I found it quite difficult to speak to what was satisfactory or unsatisfactory for my partner. It seems to me that a great deal of shielding takes place where men are concerned when it comes to sex. At least in my limited experience, having had only one partner in my life. True feelings about performance are probably not well expressed, since so much is tied up in the performance issue. Sex seems much more complicated for men, in my opinion; not physically so much as mentally and emotionally. True intimacy is a difficult thing for men to manage, in my opinion. It is what it is.
18	I have combed the internet, including Medscape, trying to learn how many couples experience a significant decline in their overall marital relationship after a dx of prostate cance. My husband was diagnosed 6 yrs ago with a gleason score of 8. Underwent IMRT and has been on luprolide, then luprolide + casedex and is now refractory with rising psa but no metastases - yet. He now is on prednison 5mg bid and after an upcoming cystoscopy to determine the cause of blood in urine, will then go on ketokonazole. I have been there for and with him every step of the way over the past 6 years. Alhtough we have talked about sex and intimacy, he has no interest, even in lying together cuddling. We now sleep in separate bedrooms. He has increasingly isolated himself from me both physically and emotionally. He is fortunate in that he feels great and has been able to work, increasing his internatioanl travel to the point where I never see him. His answer: I'm trying to outrun the cancer. We had a great relationship prior to the diagnosis, but it's been on the downward spiral ever since. I feel guilty about complaining, but I have feelings and needs as well. Do other women have these issues?
19	I have no additional comments at this time.
20	I hope that this will help others. I'm flexible, addaptable, and innovative so for me I'm satisfied with the affection, consideration, and love of a great man, sex is secondary to these...

| 21 | I think that a survey of Partners needs to be specific noting that partners are often women and this survey really doesn't specify on several of the questions. I think the bottom line is that most women I know married to prostate cancer survivors are not satisfied sexually. For most the sexual issue has been made to be let go of, kind of like, if we don't talk about it, it doesn't exist. Even when I tell my husband that I still want to have the closeness that sexual intercourse or intimacy allows for he tells me that it is only for me, that he has no feelings sexually and that sex is just mechanical, and that I just don't understand. He says that shots hurt, not just when he gets them but for the duration of his erection. The pills don't work at all. The robotic surgery, though was supposed to be nerve sparing, didn't spare anything. His physician though very updated on surgery has not a clue when it comes to a prostate cancer survivors sexuality, or soon to be lack there of. We had better sex prior to surgery than after, My husband could maintain an erection even if it was with the help of viagra/or other meds. The whole thing saddens me, our relationship is great but i think that sex would enhance it and bring us to a different level. however, his statement is that it's now only for me, my statement is that 90% of sexuality is in the mind. I guess prostate cancer has overrridden that part of it. |

22	I was in this survey because I felt my experience was important. I was with Xxxxxxx for three years. He had had RP three years before I met him. He had not dated because he was incontinent and impotent after his surgery and he had the added issue of being a kidney transplant person. In 2003, a urologist agreed to put an artificial sphincter in; and he also got caverject ! So he decided to date again. We met in March 2004 and had a wonderful relationship......he was a sensitive loving man who apparently prior to his medical issues had learned that to please a woman, there were many other things besides penises which man a woman happy. (in fact, his previous lovers seemed to never leave! I could certainly see why!) In January 2006 sadly his PSA had risen and the docs decided to radiate his body and he got the lupron shots. HIs sex drive went away and he developed terrible pain in his abdomen, probably nerve damage from the radiation. It was terrible. He was put on all this pain medication. I cant tell you how terrible it was......Oct. 21 I took him to the ER barely conscious; his lungs were a mess (I swear it was scatter from the radiation).....he ended up on a vent, and died November 15 ! A nightmare. He was only 59. I feel my experience was important. Xxxxxx was a survivor, really. Despite all these health issues, he was a wonderful, optimistic person. He took very good care of himself and realized he had to be careful. He was very sexual and accepted the caverject as a means of erection. (He had had an affair with some woman years before who taught him lots of great things about how to please a woman, and he was just great in bed.) He was loving, funny, and devoted and his death was so unexpected and premature. I hoped after the experience I would not ever deal with prostate cancer again. I met and married a wonderful man earlier this year. We learned last week that his PSA is elevated; and he needs a biopsy. It is discouraging but I am hoping that medical science is more advanced and Xxxxx will not have to go thru what Xxxxxxx (and I) went thru......we shall see.
23	I welcome the opportunity to express how badly the treatments for this disease have disturbed our lives. It goes beyond sex to general dissatisfaction with my life. It appears that my husband is doing fine - he has no interest in sex. I on the other hand feel that I aged 20 years in 3 short months. I was 59 when my husband told me he had advanced prostate cancer. He had the radical surgery with complete removal of nerves before my 60th birthday. I have felt unloved and unwanted for the last three years. I have no role except housekeeper, cook and laundress. On a good day I am able to put one foot in front of the other and get through my work day and housekeeping. On a bad day I go to bed praying that I do not have to get up in the morning. the only end to this that I see is my husband passing away from this disgusting disease and i finishing my live alone and likely impoverished by the cost of treatment.

24	I wish we had known more about post surgery issues. No matter what we read, we were truly under the assumption that in 2 years we would be fine since we had a very active sex life before. When the two years came and went, it was very discourating, especially for my partner. Most times, he just wants to give up. If it weren't for me, he would have. I have tried to tell him how important intimacy is, even if we don't have intercourse, but to him, and most men, if you dont have penetraton, you don't have sex. It has really done a number on his head and it's so sad to see. we are both divorced and only dated a year before his diagnosis. The best sex we had ever had, and now this. I try to encourage him, and do believe he can have an erection that can "work" We are now going to try daily dialis. Our first real erection was right before one year, and he had taken one cialis every other day. I think people want to know the news, good or bad and then deal. Not be hopeful and then have the air taken out of their sails. Thank you for doing this.
25	In my husband's case, HT quickly killed his libido and he had absolutely no desire for sex, including any physical "cuddling". Just always told me he wasn't interested. Very heartbreaking since we had enjoyed a very active, healthy sexual relationship for more than 30 years when the PCa treatment began. After 10+ years of such degrdation, Ino longer make any attempt to discuss this with him. It is just too emotional for me. It is so frustrating that honest questions and answers are not available to wives and partners of PCa patients right from the beginning. This is not just a man's disease...this is definitely a couple's disease and I have come to realize from taking with other women, it often impacts the woman more so than the man. How heartbreaking it is to feel like you are no longer wanted, attractive, sexual....in my mind I just became "nothing" to my husband other than someone to talk to, watch TV with and be a housemaid. It was not like that before the HT was started. I know he is very sensitive to this but he just says there is absolutely no interest......how do we live with this and and not become bitter?
26	Interesting
27	Interesting to take your survey and we'd like to be in touch - We started a blog about our sexual experience as we weren't finding information online that held true to our experience. We have great sex without erections and are alarmed at how most of the information on the web is about regaining erections while few people discuss the fact that you can actually have great sex without them. –Xxxxx
28	It has been so long since the procedure, I don't remember details very well. I was always willing to do whatever he wanted, but he quit trying for penile insertion after a few months. We satisfy each other as needed now.

29	It has not even been 3 months since my husband's cryosurgery. I want to help him in any way I can. When we go back to the uro for his 3 month follow-up, we are going to ask about the injections. I think my husband is way more upset about performance than I am. He always satisfies me. I am trying to explain about the dry orgasms and and just about relaxing and enjoying the act, even if it doesn't lead to orgasm. I guess this is easier for a woman to grasp than for a guy.
30	It makes me sad. We were very much in love. I am now 61, and haven't had sexual intercourse since before June 5, 1992.
31	Lack of success achieving penetrative sex has left us with a failing desire for sex at all. We agreed prior to his operation 8 years ago that life was more important... it seems it was so easy to believe that then. We failed to appreciate how devastating it would be to a relationship. It's like losing a loved-one; the memory is dear but the loss tangible.
32	My greatest hope is that more doctors will be told that they must address the subject of sex. If they aren't comfortable doing this, or don't have the time to do it properly, they should direct the patient & his partner to a good source of information. It's sad when the problems are not understood until they've become established, and the patient doesn't even know he could have done things to help with his rehabilitation.
33	My husband could not take stronger medication, because of other medications he was taking. We considered mechanical devices, but felt they would be unpleasant. We have been married a long time and perhaps found it easier to adjust to this deprivation than younger couples.
34	My husband had his last Lupron injection in August, it's still in his system. After reading about penile rehab for surgical cases, we wondered why the ADT guys were ignored. Even with no libido, he decided he wanted to improve his chances of recovering sexually. He first tried the VED, about 7 months after starting the Lupron. Several months later he got a script for daily Cialis - the first few mornings he had erections that a 20 year old would have loved! He still has no libido and has not had an orgasm. But, he has had a few nightime erections and we are able to have sex 2-3 mornings a week. We are cautiously optimistic about the future, even knowing that the effects of radiation are going to come into play as well.
35	My husband is depressed and thinking about suicide, he no longer feels like a complete man. He's able to please me in bed all the time and I have complete sexual satisfaction, but he has to work at it and stop several times to "pump." It sort of kills the mood and he often gives me trying.
36	My husband's surgeon never discussed sexuality other than to say that it would be entirely impossible after his surgery. This survey suggests that there are options. I would love to know more about these options and who we would get the information and treatment(s) from. Sounds like we need to do some research.

37	My partner became active in UsToo after his treatment and we have received a great deal of importaattennt information and support from that organization. Even though I do not d the regular meetings, I have come to know other members of the group and their partners through fundraisers and advocacy events.
38	My partner does not get an erection at all... he wishes he did... but he does not... not even with injections.
39	NEVER
40	Never had medical aids or assistance.
41	no comments
42	Our ordinary sex life ended with his prostate removal. I had just begun natural hormone replacement therapy and increased my libdo when he was diagnosed with prostate cancer. After I nursed him back to health, he suggested we get another sexual partner for me. I was reluctant at first, but my high libdo helped me get comfortable with the idea. I eventually turned him into my submissive, cuckold husband. We are both having a great sexual time. We get along better and are closer than ever. He gets sexual stimulation on a daily basis, achieves orgasmic satisfaction when allowed and loves watching me being pleasured by my lover.
43	Prior to my husband developing prostate cancer we had had existing intimacy issue that I never quite understood (and still don't) WHEN we had sex, it was mutually enjoyed. I always had a problem with the lack of frequency (often with stretches of 4 - 19 weeks in between!) Developping prostate cancer, and subsequent surgery thereof, has exacerbated the problem. His surgery was 18 mo. ago. while the surgeon said that the nerves were spared, and that full recovery should be expected, my husband has seen little to no improvement in the ability to achieve an errection. He HAS been able to control the leak of urine to the point that it is an almost mom-existant problem. He uses Cialis daily with no sign of improvement. I feel that his lack of attempting (either by self masturbation or by mutual masterbation) right after surgery (and since) has impeded this process. he does not feel the two are linked. I am happy to answer this survey HOWEVER would appreciate, in return, a professional 'suggestion' of where to turn for help next. thank you.
44	q. 44 We use toys!!!!! Sometimes we weren't certain about your terms.... e.g. is a toy an aid.... my husband has not had a penile implant and the viagra basically doesn't work..... we have mutual stimulation.... and orgasms utilizing our toys.... you never asked about the complicating factor of incontinence and that is a a bib obstacle for a large percentage of men who have been treated for pc. You need lots of imagination and tricks and trials to overcome this obstacle but it is possible to do it.

45	Regarding questions 44 &45, only the penile injection results in an erection so far. This makes intercourse a bit of a challenge. As yet I alone am responsible for the injection. he does not feel that he could inject himself. It feels very clinical. Consequently we postone having intercourse for lengthy periods. This is depressing for both of us. However when we connect it is extremely good. I think that going into this we were optimistic that we wouldn't lose so much. The emotional turmoil is more than I had expected. Doing this survey has been a chance to discover how I feel about the whole process.
46	Sex after radiation treatments was possible. After salvage surgery, there was no possiblilty. We've been married a very long time, are and have always been happily married. Yes, we miss sex but the libidos we once had are diminished thus making this adjustment easier. We certainly have good memories!
47	Since the hormone therapy he is just not interested. So no attempt on his behalf if I tried to instigat3e, no reaction so gave up
48	since the surgery .. errections are about a third of the size. Still works though.. He has experience leaking- dripping for first six months after radiation and surgery when errect.. caused ALOT of distress for him more than me.. now extra steps to clear the bladder are made prior to sexual activity..
49	Some of the questions speaking from the spouses perspective were not very clear. The doctor should spend more time with the couple discussing what happens after surgery. I wish there were more information on prostate cancer to help the women with this disease. It needs to be explained to men that there are a lot of different forms of foreplay, inside and outside of the bedroom. Thanks for the survey, hopefully, this will let the doctors and the patients to make better decisions.
50	Some of these questions seemed to be more geared to responses from my husband so it was a bit confusing at times. It has been a very stressful time for us and my husband blames me often for his inability to get an erection. I am not sure our marriage is going to survive.
51	Some of these questions were difficult as they do not apply to a homosexual relationship and maybe you ought to revise this survey with some sections specifically for homosexual partners as I really do not know about such things as whether orgasms are inportant to all women......etc Thank goodness, my partner got prompt treatment, has a great and proactive GP. He had surgery and there's no need for more treatment as far as we know right now. The GP assisted us making the decision ie surgery vs radiotherapy. Mt partner had surgery in Sept, about 3 months ago, so my responses need to be considered in that regard. I'm so glad he is okay and the cancer was removed.
52	Sometimes questions in this survey were geared toward the patient (exclusively) even though the spouse was completing the survey and therefore hard to answer. I think they just needed to be re-phrased.

53	Thank you for asking and doing this study. It is a very sensitive issue and needs to be talked about. It can't remain a "secret." Men and their partners need to be aware of options and possibilities.
54	Thanks for asking for input from partners. I too often hear commercials hawking products or meds that aid "performance". I say, "If it's a 'performance' I'm not too interested." My husband and I love each other deeply and whatever we do for the other is out of love and consideration--not a performance. I am happy when my husband is satisfied--I am content with affection, kind words, loving gestures. I am mindful that my admiring him for the many aspects of his personhood demonstrates to him all he means to me--nothing there is changed if things may not work like they used to. If he must go through any other procedure--hormones, radiation, etc that will change 'normal' sex--that's OK too because we have the groundwork for love and affection no matter what.
55	Thanks for asking. I think wives are left out of this equation and that the truth of how disruptive prostatectomy is to a great love relationship is glossed over with the whole, "women need hugging, kissing, lying naked" response. I believe women need a lot more than that, but have not seen this anywhere in print yet, so let me put it in print. I'm happy my husband survived surgery and cancer. I'm devastated that at 50 years old my sex life is over. Hugging and kissing is not what real women want any more than hugging and kissing is what real men want. But we women do not want to hurt our already supremely hurt men, by voicing that opinion publicly.
56	Thanks for providing a place to give information about sexuality/intimacy issues, which are core in Prostate Cancer, and yet have been very neglected in my husband's case. I was close to divorcing him before his diagnosis because of lack of emotional closeness/affection and after the diagnosis I was somewhat relieved to have what I thought was a physical reason for the problems. However, I have found that my husband does not want to discuss or deal with his sexuality and does not want me to even see him unclothed. He does not want to look at me or touch me either. Since his case is advanced, I agreed to stay with him after 42 years of marriage. His survival time with good quality of life is coming up on 5 years now, so he has done well physically. I have adjusted by moving upstairs and providing my own sexual satisfaction. I have tried many times to get him to see a therapist with me, but he refuses. I have a great psychotherapist myself, and that has made all the difference in helping me deal with him. I wish some future survey would allow both the patient and partner to elaborate more on the issue I listed above and not just use dry numbers. There are so many stories out there, many similar to mine.
57	Thank you for putting this out there for us! We need more avenues for dialogue on this issue.

58	The mind is a powerful thing-- my husband would relax in the tub and let the water wash over his genitals-- he would achieve orgasm soon after surgery without an erection. It is like exercise-- use it or lose it! With this private play, and oral sex, and MOST IMPORTANTLY-- communication, we were able to resume our healthy sex life. It may be different, and it may require planning (Viagra) but it is still highly satisfying and an integral part of our lives.
59	The sex is often satisfactory, but the leadup is very hindered. Despite what we were told, I definitely think his ED has had an effect on both his libido and mine. Spontaneous erections were one of the things that put us both in the mood. Now, needing to take a pill up to an hour or more in advance has had a very detrimental effect on spontaneity. On the other hand, if my spouse has taken a pill, I feel obligated to have sex. It's almost like his taking the pill is supposed to take the place of foreplay or intimacy to "put me in the mood." We frequently struggle with this issue. Him: Should I take a pill or not? Will she be in the mood? Her: Oh, he took a pill, I guess he'll be wanting sex. I guess I should try to get in the mood. Also, his size and the quality of erections have also been affected, which has affected my enjoyment of sex, despite the fact that my orgasms are typically not from penetration, and usually take place either before or after penetration. In other words, it stinks! One last comment about the survey: Some of the questions were confusing because they seemed to be directed at him, not me. I kept checking to see if I was taking the correct survey, e.g. the partner survey, not the survivor. But overall, glad to help with whatever information I can provide.
60	The survey is not easy to answer for me. I met my partner after prostatectomy, EBRT and his first course of androgen deprivation. We used pills and shots very successfully and his libido was good. Then he underwent a second course of ADT and IMRT, and his libido dropped to nothing, so there has been no attempt at all to have intercourse, and he is not interested in touching me at all, though he was very affectionate before. At this point, we have no intimacy and our once close relationship is very close to ending because of it. He feels since he is uninterested in sex, he is also uninterested in touching. Is supect he does not want to touch me because he thinks it will get my hopes up and he will disappoint me. There is so much to this issue, and I believe the treatment for this cancer deprives many many couples of any kind of rewarding life together.
61	There probably should have been an N/A slot for some of the questions. We were very fortunate to have been able to go to and have the nerve sparing radical procedure. At the time, NO surgeon in CT was doing that procedure. Fortunately, our insurance allowed us to go out of state. Everyone we know who had seeding or a basic radical either has sexual function issues, bladder control issues, or both. This is an important issue - should be researched.

62	These were helpful and hard to discuss with friends, except prostate cancer friends. I had a problem with trying to put down that we were together 47 years. It kept asking me for a decimal point???? I think one thing that is major and missing is that on drugs like lupron, men lose their desire as well as their ability and when we could finally sort that out (it took months of anguish for both of us), we have been able to move on in a different direction. The thing that finally clicked for my husband was a 3rd or 4th reading of Patrick Walsh's description of the grief for long term partners as this moves along AND my friend whose husband is on lupron. These issues were never discussed with our physician who we really like, but who just was not able to have that discussion, except to say it would destroy the testosterone. I guess neither of us knew about that the desire piece was SO related to testosterone. I hope other couples get that information ahead of this medication. Thank you.
63	This has been a great frustration to me. I am not sure our relationship is going to survive this. I don't feel my partner really cares to work on a solution. He seems to have completely lost his sex drive. He doesn't follow the advice given to him by his dr. Or medical team. It is very disappointing to me.
64	This has been a very difficult time for me. The facts that his cancer had progressed into the seminal vesicles and the nerves were taken made me heartsick. Worried that the cancer can come back and that we could never again have mutually satisfying intercourse, I cried a lot. My husband was in more pain than he ever had been before but I was emotionally centered on what we may have lost. But, when we got home, with the catheter removed, we were able to communicate better and to try to soothe each other's struggles. We both stimulated each other orally. He wanted to allay my worries and please me. I wanted to heal him and encourage blood flow. He drank the tears I cried. Before the surgery, the doctor briefly touched on ED, just saying that it could always be fixed. When my husband looked at me sadly, he said, "She'd rather have you." How could I disagree? But after the surgery, I was disconsolate and angry. I wanted him in me, not just for orgasm but for the union and bonding. After I spoke to the doctor on the phone, he sent us a vacuum devise. It was wonderful and disappointing. Since then, we've both received more pleasure from it's use. But, we continue to work towards eventual natural erections. The doctor has been more informative and helpful since and I'm more at peace. My husband is taking Viagra regularly and will be learning to give himself injections. Just in case some nerves are left, we want to improve the blood flow. We're more in love than ever.
65	This survey does not seem to recognize that in androgen deprivation therapy a lack of libido is often the problem.

66	This survey is appropriate for a person who has dealt with 1st line of treatment for prostate cancer. The questions are very general, and do not allow for the progression of disease and the effects of changing treatments from progression on erectile function/dysfunction. As in our story, each treatment had a different effect on my husband's ability to maintain an erection. Our emphasis on our sexuality as a couple had to change to meet his ability with each change from treatment. There aren't any questions that address the man's lack of INTEREST in sexuality as a result of hormone deprivation therapy and the impact this has on a couples sexuality. This topic seems to be missing in many discussions on sexuality and prostate cancer with the focus more on the ability to have an erection. I would encourage more future discussion/research on this topic. I am glad you are pursuing this very important topic. In our support group we often emphasize that prostate cancer is a disease that impacts both the individual and their partner.
67	Trimix 1st solution dose was not enough & unsuccessful, the 2nd mix is better & caused an erection. Theres an adjustment period to planning-taking a shot, takes some romance out of it. There is different strengths of Trimix. The surgeon didn't discuss what the effects would be afterwards, the Urologist has given us our sexual life back & was very informative. Family doctor was helpful in referring us to him. No one else, even the counsellor did not provide us with the facts and high percent that men after surgery will not be able to have a spontaneous erection after surgery & that treatment should be started asap.
68	We were starting to practice sex, cuddling etc. again when my husband found out that his PC may be returning - ie elevating PSA. I too have had health problems - an unruptured brain aneurism, so sex has not been at the forefront of our lives. Its a pity but I do think that probably we need to get our equilibrium back after all the health shocks we have been given over the past months, then maybe we can get some joy back.
69	We've managed to maintain a fairly active sex life, albeit, all oral sex. We miss the penetration, intimacy from penetration, although I've never had a vaginal orgasm. I've had many pleasurable clitoral orgasms. Partner education/support is needed, at least resourses provided by physicians. Thanks for your research!
70	When we made the treatment decision; intercourse was important to us. By the time his treatment failed I was in menopause and no longer had a libido. By the time he began hormonal therapy we were both content to no longer need intercouse. Now we share the hot flash experience and more or less "grin and bear it". Funny how metastatic cancer changes your views on what is and is not important in a relationship.
71	With regard to 44 and 45, I answered the question referring to the first 24 months. After that, no aids have been necessary and my partner has had successful erections almost all the time with satisfaction for both us of.

In reviewing these comments, we remarked on the amount of good, helpful information that is available. Good information could help to eliminate many of the problems dealing with erectile function, and alleviate much sadness, despair and frustration.

35% of the partners took the time to comment. The information we gained from both survivors and partners has been enough to justify the whole survey. Your openness and honesty indicate the need for more and better quality information, and we will do our best to honor your trust.

We cannot overemphasize how helpful the comments have already been for many survivors and their partners.

In closing *some of the more important findings of the study are summarized below.*

As the survey dealt with issues of post treatment sexuality, we asked the participants about their abilities to perform penetrative sex and their feelings about being able to reach an orgasm, a question that has rarely been asked in this type of survey.

Ability to perform penetrative sex:

classified by treatment and time

	@ 3 months	@ 24 months	Change over time
Prostatectomy	16% able	28% able	12% improved
Radiation	67% able	40% able	27% regressed

Importance of Orgasm:

- *Unimportant: A mere 8% of both the survivors and partners felt it was unimportant.*
- *Important to Very Important: 60% of the survivors with concurrence of 42% of their partners*

As there was a significant erectile dysfunction amongst the respondents, would it be beneficial to explore whether orgasm could be substituted for penetrative sex?

It was abundantly clear that prostate survivors and their partners need assistance in dealing with the sexuality issues and the professionals in the field need to develop timely assistance and more support for prostate cancer survivors and their partners.

We will do our utmost to honor your efforts and get this information out to those who will benefit from it.

We hope that this information will continue to be a resource for everyone dealing with prostate cancer

Resources & References

Bibliography

1.   Bachunas AB.Best practices for Post-Prostatectomy Penile Rehabilitation/ED. Treatment and Urology Incontinence. http://wjweiser.com/securesite/uapa/meetings/2012/2012%20UAPA%201st%20Annual%20Meeting%2001.20.12-01.22.12/Friday/pdfs/Anthony%20Balchunas%20Best%20Practices%20for%20Post-Prostatectomy%20Penile%20Rehabilitation.UAPA1.pdf

2   Haroch G. Whats Up? Early intervention is the key to penile rehabilitation after prostatectomy. Our Voice Vol27 No4www.ourvoiceinprostatehealth.com/articles/2012/12/what'sip.

3   Dall'Era J E, Mills HK, Koul H K, Meacham R B. Penile rehabilitation after radical prostatectomy. Important therapy or wishful thinking? Rev Urology. 2006 Fall; 8(4):209-215

Internet sites cited by Prostate Cancer Survivors: survey respondents.

1. Ustoo.org
2. Prostatepointers.org/mailman/listinfo/p2p (Patient to Physician)
3. Prostatepointers.org/mailman/listinfo/circle (support for partners)
4. Prostatepointers.org/mailman/listinfo/seedpods (those interested in Brachytherapy)
5. Prostatepointers.org/mailman/listinfo/pcai (mailing list on intimacy)
6. Prostatecancerinfolink.net (the new PCa infolink)
7. Prostatecancerinfolink.ning.com
8. Prostatecancerinfolink.com
9. Prostatepointers.org (division of UStoo)
10. Pcainaz.org (Arizona branch of UStoo)
11. NIH.com (enter prostate cancer)
12. Prostatecancer.ca (Canadian PCa website)
13. Peiprostatecancersupport.com (Prince Edward Island, Canada support group)
14. Facebook has groups for PCa survivors
15. Healing well.com
16. Amedeo.com (Medical literature guide)
17. Prostatevideos.com (Dr. Gerald Chodak)
18. Netzero.org (in search box enter prostate cancer)
19. Protonbob.com (for proton beam news letter)
20. Yananow.com (lists PCa sites)